AN EVALUATION OF WORLD BANK SUPPORT, 1997–2007

Water and Development
Volume 1

2010
The World Bank
Washington, D.C.

Cover photo: The Bund, Shanghai, China. Two children along the Bund admire the Shanghai skyline. Photo courtesy of Jody Cobb/Getty Images.

ISBN: 978-0-8213-8393-3
eISBN: 978-0-8213-8394-0
DOI: 10.1596/978-0-8213-8393-3

Library of Congress Cataloging-in-Publication data have been applied for.

World Bank InfoShop
E-mail: pic@worldbank.org
Telephone: 202-458-5454
Facsimile: 202-522-1500

Independent Evaluation Group
Communications, Learning, and Strategy
E-mail: ieg@worldbank.org
Telephone: 202-458-4497
Facsimile: 202-522-3125

Printed on Recycled Paper

Table of Contents

Boxes

Figures

Tables

Volume 2: Appendixes

Find volume 2 online at: http://www.worldbank.org/ieg/

Abbreviations

Bibliography

BP	Bank Procedure
BWO	Basin water offices
CDD	Community-driven development
DFID	Department for International Development (U.K.)
DPL	Development Policy Loan/Lending
DSM	Demand-side management
EFA	Environmental flow assessment
ERR	Economic rate of return
ET	Evapotranspiration
GEF	Global Environment Facility
GW	Gigawatt
IBNET	International Benchmarking Network for Water and Sanitation Utilities
IBRD	International Bank for Reconstruction and Development
ICR	Implementation Completion Report
IDA	International Development Association
IEG	Independent Evaluation Group
IFC	International Finance Corporation
IMF	International Monetary Fund
IPCC	Intergovernmental Panel on Climate Change
IWMI	International Water Management Institute
IWRM	Integrated water resources management
MARPOL	International Convention for the Prevention of Pollution from Ships
MCIPR	Mid-Cycle Implementation Progress Report
MDG	Millennium Development Goal
MIGA	Multilateral Investment Guarantee Agency
MNA	Middle East and North Africa
MPA	Marine protected area
NBI	Nile Basin Initiative
NILE-COM	Nile Council of Ministers
O&M	Operation and maintenance
OP	Operational Policy
PSP	Private sector participation
RBO	River basin organization
SMT	Social mobilization
UfW	Unaccounted-for water
UN	United Nations
UNDP	United Nations Development Programme
UNEP	United Nations Environment Programme
UNICEF	United Nations Children's Fund
WHO	World Health Organization
WPI	Water Poverty Index
WSS	Water supply and sanitation
WUA	Water user association

Acknowledgments

This evaluation of the World Bank's experience with the multiple aspects of water was done by the Independent Evaluation Group–World Bank at the request of the Bank's Board of Executive Directors.

The evaluation was conducted under the leadership of Ronald S. Parker, and this report was written by Ronald Parker with research support from Anna Amato, Gunhild Berg, Carola Borja, Ahmad Hamidov, Elizabeth Eggleston Helton, Silke Heuser, Kristin Little, and Emily O'Sullivan. Silke Heuser provided methodological and data analysis support to the team. Gunhild Berg, Silke Heuser, and Kristin Little summarized the background documents and wrote sections of this report. William Hurlbut edited the report and provided editorial support for the preparation of background papers and case studies. Michael Treadway and Caroline McEuen edited the volumes for print and Web publication. Marie Charles provided administrative support. Antoine Boussard assisted with the fieldwork for the country case studies. Keith Pitman provided comments on selected background papers. Peer reviewers Keith Oblitas and Meike van Ginneken provided valuable comments on earlier drafts of the report. The authors gratefully acknowledge the members of the External Advisory Panel—Mohamed Ait Kadi, Mary Anderson, Judith Rees, and Peter Rogers—for their early guidance and their patient and detailed review of drafts of this report, and the Lincoln Institute and Gregory K. Ingram for making their facilities available to us for our first Advisory Panel meeting. The statement of the Advisory Panel is included in this volume.

We also gratefully acknowledge the financial support provided by our partner, the Swiss Agency for Development and Cooperation (SDC), which made possible a considerable amount of the background research upon which this evaluation is based. We also thank Abel Mejia for continual advice and encouragement. The study carried out several field missions. The authors thank Bakyt Arystanov, Gabriel Azevedo, Alain Barbu, Sabah Bencheqroun, Francis Ato Brown, Rafik Fatehali Hirjim, Ospan Karimsakov, Jerson Kelman, Nurlan Kemelbekov, Dilshod Khidirov, Hassan Lamrani, Washington Mutayoba, Phillipo Patrick Mwakabengele, Mohamed Oubelkace, Jennifer Sara, Tauno Skytta, and Huong Thu Vu for their support and valuable insights during these missions. A database of project information created by Anna Amato for this evaluation brought together for the first time all the available information on every Bank project that had water-related activities. This database will be turned over to the Water Anchor for use by Bank water staff.

Director-General, Evaluation: *Vinod Thomas*
Director, Independent Evaluation Group–World Bank: *Cheryl W. Gray*
Manager, IEG Sector Evaluation: *Monika Huppi*
Task Manager: *Ronald Parker*

Only 1 percent of the world's freshwater is available for human use, and the amount of available water has been constant for millennia. Meanwhile, the planet has added 6 billion people. Vastly uneven distribution of freshwater resources, combined with changes due to climate change, is already deepening water-related problems.

Development patterns, increasing population pressure, and the demand for better livelihoods across the globe will all contribute to the global water crisis. Addressing that crisis will require maintaining a sustainable relationship between water and development, one that balances current needs against the prospects for future generations.

The Bank's strategy for water, together with countries and partners, has made an important difference. However, the approach taken thus far has underemphasized some of the most difficult challenges, which in turn has left some crucial needs unmet. Moreover, the Bank's involvement in water will face heightened challenges due to climate change, migration of people to coastal zones, and the declining quality of the water resources available to most major cities and industry. These will require important adjustments to the approach followed. Growing water scarcity is a reality, which the Bank and its partners need to confront by putting more emphasis on the challenging areas of groundwater conservation, pol-

Photo courtesy of Curt Camemark/World Bank.

Water has been a major focal area for World Bank lending to developing countries. Almost a third of all Bank projects approved since 1997 have been water related. Water loan commitments grew 55 percent during fiscal 1997 to end-2007.

During those 11 years, project performance improved steadily against stated objectives, led by a notable project performance improvement in Africa. Water also has been integrated into many other sectors. The Bank has contributed to improving access to clean water, especially in urban areas. On the institutional side, it has balanced investments in infrastructure with improvements in the institutions that manage and allocate water.

lution reduction, and effective demand management. New ways need to be found to help the most water stressed countries make water sustainability a cornerstone of their development plans. The development community also needs to help countries shift more attention to sanitation.

Ever more powerful storms and rising sea levels threaten increasingly densely settled coasts. This situation calls for more strategic development planning and more effective disaster risk reduction for low-lying coastal areas. Finally, data collection and use need to be enhanced in a number of areas. In all of these activities, strong partnerships and knowledge creation and sharing will continue to be essential.

Vinod Thomas

Vinod Thomas
Director-General, Evaluation

The amount of available water has been constant for millennia, but over time the planet has added 6 billion people. Water is essential to human life and enterprise, and the increasing strains on available water resources threaten the mission of institutions dedicated to economic development. The ultimate goal is to achieve a sustainable balance between the resources available and the societal requirement for water.

In this evaluation the Independent Evaluation Group (IEG) examines all the water-related projects financed by the World Bank between fiscal 1997 and the end of calendar 2007. Bank activities related to water are large, growing, and integrated. They include water resources management, water supply and sanitation, and activities related to agricultural water, industrial water, energy generation, and water in the environment.

Changes over the decade in the Bank's portfolio of water projects have been broadly positive. In 1997 only 47 countries borrowed for water, but by 2007 there were 79 borrowers, and lending for water had increased by over 50 percent. Water projects have had good success rates relative to their goals, and this performance improved in the latter half of the evaluation period—with a particularly notable 23-percentage-point improvement in Africa. Within the Bank, water-related activities have been supported by institutional changes, and there has been progress in integrating water into the work of other sectors.

At the same time, against emerging problems and pressures, crucial needs remain unmet. In the complex area of water resources management, it has often been easier to underemphasize the most difficult problems, such as fighting pollution or restoring the environment, compared with such tasks as purchasing equipment or building infrastructure. Limited success with full cost recovery for water services has caused the Bank to moderate its approach, but the question of who will pay for uncovered costs remains to be resolved.

With borrowers facing increasingly difficult challenges in water management, business as usual is not an option. The evaluation suggests that the Bank and its partners should find ways to support systematically the countries that face the most water stress. It recommends that more attention be given to critical concerns of groundwater conservation, pollution reduction, and coastal management and that the

Bank work with clients to shift more attention to sanitation. Demand management must be a theme of Bank support if the challenges of increasingly scarce water are to be tackled successfully, and the Bank and its borrowers need to take a clear stand on cost recovery. Finally, data collection and use need to be enhanced in a number of areas. In all of these activities, strong partnerships and knowledge creation and sharing will continue to be essential.

Only 3 percent of the world's water supply is freshwater, and two-thirds of that is locked in glacier ice or buried in deep underground aquifers, leaving only 1 percent readily available for human use. Water is not only limited, but unevenly distributed. In more arid regions, water shortages are always a threat. Moreover, the scientific consensus is that climate change will worsen these water-related challenges in the coming years. These changes are already disrupting rainfall patterns, feeding ever more powerful windstorms, and creating droughts of unprecedented severity and frequency. About 700 million people in 43 countries are under water stress.

Development patterns, increasing population pressure, and the demand for better livelihoods in many parts of the globe all contribute to a steadily deepening global water crisis. Development redirects, consumes, and pollutes water. It also causes changes in the state of natural water reservoirs—directly, by draining aquifers, and indirectly, by melting glaciers and the polar ice caps. Maintaining a sustainable relationship between water and development requires that current needs be balanced against the needs of future generations.

The development community has transformed and broadened its approach to water since the 1980s. As stresses on the quality and availability of water have increased, donors have begun to move toward more comprehensive approaches that seek to integrate water into development in other sectors.

Through both lending and grants, the World Bank (the International Development Association and the International Bank for Reconstruction and Development) has supported countries in many water-related sectors. This evaluation examines the full scope of that support over the period from fiscal 1997 to the end of calendar 2007. More than 30 background studies prepared for the evaluation have analyzed Bank lending by thematic area and by activity type.

The evaluation is by definition retrospective, but it identifies changes that will be necessary going forward, including those related to strengthening country-level institutions and increasing financial sustainability.

Water and the World Bank

The Bank's 1993 Water Resources Management Policy Paper moved the institution away from its previous focus on infrastructure development for the water sector. The paper also shifted the Bank's planning process from one based on discrete investments within the sector to a multisectoral approach, embracing the concept of integrated water resources management (IWRM). IWRM promotes the coordinated development and management of water, land, and related resources in order to maximize economic and social welfare in an equitable manner without compromising the sustainability of vital ecosystems. Under IWRM, each water-related activity in a project or program is considered carefully in light of other competing uses and its social, economic, and environmental consequences.

In 2003 the Bank adopted a new water resources strategy (World Bank 2003b) that looked more closely at water management and the connections between resource use and service delivery. It also reintroduced infrastructure investments as an important aspect of Bank support in the sector. The 1993 and 2003 strategy papers are complementary, and together with the Bank's mandate to reduce poverty, they have helped inform issues of supply and improve the performance of utilities and user associations. The 2003 strategy committed the institution to facing the most pressing challenges that were constraining the achievement of goals set in 1993.

The Water Portfolio

A large part of what the Bank finances has something to do with water: 31 percent of all Bank projects approved since 1997 are related to water. Between fiscal 1997 and the end of calendar 2007 the Bank approved or completed 1,864 projects with at least one water-related activity. Together, these projects represented Bank financing of about $118.5 billion, of which $54.3 billion was directed to water. The average loan was for $67 million (exclusive of grants and nonlending activities).

Many of the Bank's water-related activities are integrated into projects doing other things, such as developing water supply in an urban services project or drafting water policy within a larger environmental policy framework. The largest activity categories by number of projects are those dealing with wastewater treatment and irrigation. The largest amounts of money have gone to projects that involve irrigation and hydropower or dam activities.

The Bank engaged 142 countries in lending for water during the evaluation period. Of these, the top 10 accounted for 579 projects (31 percent) covering 56 percent of total Bank commitments for projects with water-related activities (nearly 5 percentage points more than those countries' share of Bank lending as a whole). China, the single largest borrower for water projects, accounted for 16 percent of water-related lending, but only 7 percent of total Bank lending.

Main Findings

Increased Lending and Improving Project Performance

The Bank increased its lending for water and the number of countries served during the period evaluated. Although the number of countries that borrow for water projects has varied from year to year, 79 countries were served in 2007, compared with 47 in 1997. Lending for water increased by over 50 percent during the period.

The integration of water practice across Bank sectors appears to be well under way. Integration of the Bank's water practice was an important goal of the 2003 water strategy, and during the period evaluated, the majority of water-focused projects were overseen by sector boards other than the Water Supply and Sanitation Sector Board.

Water projects in the aggregate have good success rates when measured against objectives. IEG performance ratings show steady improvement in the sector's performance measured against project objectives. During the most recent five-year period, water was the most improved major sector by this criterion, with a particularly noteworthy 23-percentage-point improvement in the share of satisfactory projects undertaken by the Africa Region. Within the portfolio, 77 percent of the 857 completed projects had an aggregate outcome rating of moderately satisfactory or better, slightly above the Bank-wide average of 75 percent. The trend continued in 2008, in which year water sector projects attained a 90 percent satisfactory rate.

The focus of Bank activity within the water sector has shifted over time. The Bank has lent heavily for irrigation and water supply, and dams and hydropower have become more important in the last few years. But some activities that are of growing importance as water stress increases have become less prominent in the Bank's portfolio; notably, these include coastal zone management, pollution control, and to a lesser degree groundwater conservation. Although the portfolio has performed well when measured against projects' stated objectives, the Bank and the borrowing countries have not yet sufficiently tackled several tough but vital issues, among them broadening access to sanitation, fighting pollution, restoring degraded aquatic environments, monitoring and data collection, and cost recovery. Where it has lent for hydrological and meteorological monitoring, the Bank has focused on providing technology for data collection and relatively less on gathering and interpreting information for which there is an identified demand. Such aggregate findings, however, mask Regional and country-specific variations and needs. For example, the East Asia and Pacific and Africa Regions have responded more actively than other Regions to the sanitation challenge. These issues are covered in greater detail below.

Water Resources Management

Effective demand management is one of several critical challenges worldwide in the face of increasing water scarcity. Demand for water can be affected by three broad sets of measures: pricing, quotas, and measures to improve water use efficiency.

Efforts to improve the efficiency of water use and limit demand in the agriculture sector, the largest consumer of water, have had limited success. Efficiency-enhancing technologies alone do not necessarily reduce the use of water on farms, and efforts to manage demand by charging agricultural users for water have had limited success, partly because of the low price elasticity of that demand. Fixing and enforcing quotas for water use is a relatively recent approach and deserves careful evaluation after more projects featuring this approach have been completed. Cost recovery in Bank-supported projects has rarely been successful: only 15 percent of projects that attempted cost recovery achieved their goal. Those that have succeeded have generally improved the efficiency of water institutions at collecting fees. This limited success has caused the Bank to moderate its approach, but as it has yet to clearly identify alternative sources to finance the recovery shortfall, the sustainability of investments is threatened.

In the area of water supply, reducing unaccounted-for water (UfW) has been the main activity directed at improving water use efficiency. About half of projects that attempted to address UfW managed to reduce it by at least 1 percent.

Finding effective ways to improve water use efficiency and manage demand for water will be critical if the Bank wants to maintain a leading role in this area.

Integrated water resources management, the focus of two consecutive water strategies, has gained traction within the Bank, but has made limited progress in most client countries. Within the Bank there has been considerable progress in integrating water into the work of other sectors and in consolidating institutional structures to carry out water-related activities. However, outside the Bank, even in countries where IWRM is now well integrated into the legal framework, it is known mainly in the water sector. The information necessary to inform decision making

is not easily available, and, perhaps more important, the economic implications of water constraints are not widely appreciated. Meanwhile, there are indications that the Bank is paying less attention to data collection—an essential prerequisite for successful IWRM implementation, because countries have less motivation to confront a situation with unknown parameters.

Where IWRM has been successful, it has most often been in a particular location at a time of necessity. Some countries have made progress with water resources management after natural disasters, for example. Such shocks often do not affect entire countries, however, nor are they a desirable route to IWRM. The way to open the window of opportunity without waiting for a calamity is to support monitoring processes that deliver information to relevant public and private stakeholders. The example of Brazil shows that making water data publicly available over the Internet helps increase stakeholder concern, which in turn helps to mobilize the political will necessary to confront entrenched water problems.

The number of projects dealing with groundwater issues has been declining, although within that problematic trend the portfolio has also witnessed a positive shift away from a focus on extraction. This shift is important given falling water levels in critical aquifers in many Bank borrowers.

Within the groundwater portfolio, activities aiming to increase water supply were, as a group, the most successful, whereas activities related to reducing pressure on groundwater, and to conservation, generally proved more challenging. Yet such activities will need to become more prominent in the portfolio, if the Bank is to effectively help the growing number of water-stressed countries address increasing groundwater scarcity. In the Republic of Yemen, for example, improved tube well technology and generous subsidies on diesel fuel have led to rapidly rising consumption of water for irrigation, with the result that irrigation now extracts over 150 percent of the country's renewable water resources.

Watershed management projects that take a livelihood-focused approach perform better than those that do not. Projects combining livelihood interventions (that is, the creation of income-generating opportunities) with environmental restoration enjoyed high success rates, but the effects on downstream communities (such as reduced flooding and improved water availability) and the social benefits in both upstream and downstream communities were often not measured. Hydrological monitoring (with or without remote sensing) and watershed modeling could help improve impact assessment and thus make it easier to estimate the cost-benefit ratio of such interventions.

Environment and Water

Environmental restoration has been underemphasized in the Bank's water portfolio, possibly because its immediate and long-term financial importance is unclear. More attention to cost-benefit calculations could help the Bank and its clients evaluate trade-offs and get better results.

Most Bank water projects focus on infrastructure, even though in some cases environmental restoration is more strategically important. It is not always necessary to restore the water-related environment to a pristine state in order to obtain major social, economic, and environmental benefits and reduce vulnerability. Priority improvements to degraded environments, even when small, can have big impacts. A coastal wetlands protection project in Vietnam, for example, successfully balanced reforestation with livelihood needs. The project successfully reforested critical areas and led to a substantial reduction in coastal zone erosion.

Countries and donors will need to focus more on coastal management, because some 75 percent of the world's population will soon be living near the coast, putting them at heightened risk from the consequences of climate change. Approvals of Bank projects in this area have dwindled over time, and the reasons for this should be considered in the Mid-Cycle Implementation Progress Report.

Many projects contain funding for water quality management, but few countries measure water quality. The number of projects that actually measure water quality is declining. Evidence of improved water quality is rare, as are indications of the improved health of project beneficiaries. The data that are generated need better quality control. Water quality in the top five borrowing countries is declining, and fewer than half of projects that set out to monitor water quality were able to show any improvement.

Water Use and Service Delivery

The Bank has increasingly focused on water service delivery, but there has been a declining emphasis on monitoring economic returns, water quality, and health outcomes. Only a third of wastewater treatment and sanitation projects calculated economic benefits.

Sanitation needs greater attention. Population growth in developing countries has been rapid, as has urbanization. An expansion of piped water services and increased household water use will accelerate demand for adequate sanitation. The evaluation recognizes that even if the Millennium Development Goals (MDGs) for clean water supply are achieved, 800 million people will still lack access to safe drinking water in 2015, but many more—1.8 billion—will still lack access to basic sanitation. Within sanitation projects, more emphasis is needed on household connections. Connection targets in projects are generally not met, and IEG has seen a number of treatment plants functioning below design capacity because households have not connected to the systems, in part because willingness to pay has been overestimated and facilities have been overdesigned. This report highlights the particular weakness of sanitation institutions, which will continue to constrain progress until their capacities improve.

Hydropower projects have performed well, and significant untapped potential remains for appropriate development, particularly in Africa. After a peak in the mid-1990s, dam construction in the developing world slowed. The Bank has recently increased its financing for dam construction, in many cases for multipurpose dams that provide hydropower and often also support irrigation, flood protection, or industrial use. Almost a third (66) of the 211 Bank-financed dam and hydropower projects covered in the evaluation rightly focused on dam rehabilitation, as many dams have experienced gradual deterioration brought about by lack of maintenance, and a number have been shut down because of salinity, sedimentation, and other problems. A new hydropower development business plan, "Directions in Hydropower" (World Bank 2009), was completed in 2009 and supports feasibility studies so that projects will be technically, economically, and environmentally appropriate. Indeed, it will be vital to take on board the experience with hydropower projects, including their scale, socioeconomic, and environmental impacts.

Institutions and Water

Water services are delivered by public providers in most countries, although private sector participation has made some progress. Where international private firms have been successful at providing water services in urban areas, they have contributed significant investments to infrastructure and in some cities have managed to increase the efficiency of water utilities' operations. In some Bank-financed projects in rural areas, in contrast, the local private sector manages the operation of water systems but has invested little and shared little of the financial risk. Where governments want private involvement, a well-functioning, well-maintained regulatory system is necessary for its sustainable participation in utility operations. In many cases such a system has remained elusive, and this has limited private sector involvement.

Water projects operating in a decentralized environment have had difficulty meeting expectations, but when the budget and authority accorded to the lower level of government have matched the responsibility assigned to it, projects have had positive achievements. Half of projects that aimed to strengthen local capacity and two-fifths of projects that supported institutional reforms were successful. Other positive outcomes usually associated with decentralization—increased accountability, ownership, empowerment, and social cohesion—were achieved in a minority of cases.

Support for institutional reform and capacity building has had limited success in the water sector. Institutional reform, institutional strengthening, and capacity building have been the activities most frequently funded by Bank water-related lending. Yet these interventions have often been less than fully effective, and weak institutions have often been responsible for project shortcomings.

The Bank has been actively engaged in addressing transboundary water issues. Priority has been given to projects serving waterways shared by a large number of countries. Here the Bank has been more successful in helping to address disputes than in strengthening transboundary institutions. Its work with borrowers on transboundary aquifers is in its early stages.

Strategic Issues

The Bank's complementary strategies for the water sector have been broadly appropriate. However, their application thus far has underemphasized some of the most difficult challenges set by the 2003 strategy, and this has left some needs unmet. The Bank's approach to water will face heightened challenges brought about by climate change, migration to coastal zones, and the declining quality of the water resources available to most major cities and industry in the coming decades. These will require some shifts in emphasis.

Water stress needs to be confronted systematically. At present there is no statistical relationship between Bank water-related lending to countries and the degree of water stress in those countries. The issue for the Bank is how to find an entry point and help the most water-stressed countries put the pieces together so that water needs can become more central to their development strategy. This is not to say that the Bank should stop providing support to water-

rich countries, nor is increasing lending to water-stressed countries the only or even necessarily the best solution. The failure to meet human needs for water and sanitation has its roots in political, economic, social, and environmental issues. These are becoming more entwined and cannot be solved unless a broader range of actors gets involved.

The most water-stressed group consists of 45 countries (35 of them in Africa) that are not only water poor but also economically poor. Country Water Resource Assistance Strategies have helped to place water resource discussions more firmly in the context of economic development in the countries where they have been done. Including ministries of planning and finance in the dialogue is another critical step, as is expanding the calculation of economic benefits to increase countries' understanding of the economic importance of water.

Collaboration with other partners is particularly important, and it is likely to increase in importance as the Bank helps countries tackle water crises. This is true not only for water supply and sanitation but also for water resources management in national and transboundary basins. Many of the problems described in this report are far too big for the Bank to tackle on its own.

Successful implementation of the Bank's Water Resources Sector Strategy will require a great deal of data on water resources, and therefore data gathering must become a higher priority. Data on all aspects of water and on relevant socioeconomic conditions need to be more systematically collected and monitored. Data need to be used better within projects. For example, the collection and analysis of up-to-date groundwater data are more important now than ever and need to be taken on board more commonly than they have been.

Recommendations

- Work with clients and partners to ensure that critical water issues are adequately addressed.

  - Seek ways to support the countries that face the greatest water stress. The Mid-Cycle Implementation Progress Report should suggest a way to package tailored measures to help the Bank and other donors work with these clients to address the most urgent needs, which will be far more challenging as water supply becomes increasingly constrained in arid areas.

  - Ensure that projects pay adequate attention to conserving groundwater and ensuring that the quantity extracted is sustainable.

  - Find effective ways to help countries address coastal management issues.

- Help countries strengthen attention to sanitation.

- Strengthen the supply and use of data on water to better understand the linkages among water, economic development, and project achievement.

  - In project appraisal documents, routinely quantify the benefits of wastewater treatment, health improvements, and environmental restoration.

  - Support more frequent and more thorough water monitoring of all sorts in client countries, particularly the most vulnerable ones, and help ensure that countries treat monitoring data as a public good and make them broadly available.

  - In the design of water resources management projects that support hydrological and meteorological monitoring systems, pay close attention to stakeholder participation, maintenance, and the appropriate choice of monitoring equipment and facilities.

  - Systematically analyze whether environmental restoration will be essential for water-related objectives to be met in a particular setting.

- Monitor demand-management approaches to identify which aspects are working or not working, and build on these lessons of experience.

  - Clarify how to cover the cost of water service delivery in the absence of full cost recovery. To the extent that borrowers must cover the cost of water services out of general revenue, share the lessons of international experience with them so they can allocate costs most effectively.

  - Identify ways to more effectively use fees and tariffs to reduce water consumption.

  - Carefully monitor and evaluate the experience with quotas as a means to moderate agricultural water use.

Management welcomes this evaluation of World Bank support for water, covering the period 1997 through 2007, by the Independent Evaluation Group (IEG). This evaluation is timely against the backdrop of the 2003 Water Resources Sector Strategy Mid-Cycle Implementation Progress Report under preparation; it is also relevant in view of the large and increasing volume of water lending, which now represents 10 percent of the World Bank's portfolio.

IEG's comprehensive review shows that the World Bank has been engaged in IDA and IBRD countries across the whole spectrum of water issues—from floods and droughts to rivers, lakes, wetlands, and groundwater aquifers; from access to hydropower energy to the achievement of the water supply and sanitation Millennium Development Goals; from effective demand management to irrigation and drainage, and on to cooperation through water-sharing arrangements among riparian states. Notwithstanding the achievements in these areas noted by IEG, the challenges in the sector are still significant: poor governance, financial under-recovery, intermittent supplies, growing water scarcity, and deteriorating water quality are some of the issues that are tackled as part of the Bank's regular design of water projects. Water is a complex sector; it is political in nature, and it impacts on many vital sectors of the economy, including agriculture, energy and environment, and health.

Management welcomes the overall positive findings from the review. Among the important elements of this evaluation are (a) its emphasis on the centrality of water for the sustainable development agenda; (b) the assessment that the strategies outlined in the 2003 Bank Water Resources Sector Strategy and the 1993 Bank Water Resource Management Policy Paper have been broadly appropriate; (c) the recognition that achievements have been made under each of the objectives of these strategies; and (d) the need, as the strategies are further implemented, for the development community and client countries to heighten attention to certain areas, such as coastal management, groundwater, sanitation, and data collection.

That said, management is of the view that the IEG review might have gone further, in terms of widening the coverage of water-related activities beyond project financing and providing more specificity to its recommendations. Management would also like to clarify its position vis-à-vis cost recovery versus full cost recovery, where there may be differences with IEG. The comments below on the analysis of the review reflect this differing view.

Management's specific responses to IEG's recommendations, with which it generally agrees, are noted in the attached Management Action Record.

Management Comments

Evidence shows that achieving *full* cost recovery in water services delivery is an ultimate goal, which although desirable economically is difficult and rarely achieved in practice. Underpricing of water supply services is widespread, even in upper-middle-income countries and high-income countries. Globally, estimates show that 39 percent of water utilities have average tariffs that are set too low to cover basic operation and maintenance (O&M) costs. A further 30 percent have tariffs that are set below the level required to make any contribution toward the recovery of capital costs. Even in high-income countries, only 50 percent of water utilities charge tariffs high enough to cover O&M costs and partial capital recovery. Some degree of general subsidy is thus the norm, even in high-income countries. In Bank client countries, low tariffs (that is, below full cost-recovery levels) ensure that water services are affordable to the population. While raising tariffs to recover a greater share of costs in order to mobilize private financing, or simply reducing the use of scarce fiscal resources by utilities, may be economically sound, political constituencies have often prevented tariffs from being increased. Some estimates suggest that water tariffs may have to increase by 90 percent in some developing countries to achieve full cost recovery. In the discussion about affordability, there is also a particular concern for the poor, who are disproportionately impacted by increased tariffs.

Cost Recovery in Bank Water Projects

Cost recovery continues to be central to the design of Bank water services projects. Through a series of projects, the Bank has supported government efforts to move water utilities through the continuum of cost recovery, starting with covering O&M costs. Other options to reduce the costs of services delivery are also considered in Bank projects, including the use of alternative technologies, differentiated service levels, and flexibility in standards.

Other Initiatives to Address the Sustainability of Water Service Delivery. The Bank is continuing to work on several fronts toward finding a sustainable solution to water services delivery, while increasing water coverage. The Bank has proactively examined ways to address the question of who will pay for uncovered costs. The water sector now features prominently in Public Expenditure Reviews, with a view to identifying ways to increase the effectiveness of overall public spending as well as water-specific public spending. More attention is also devoted to the transparency of water sector financing through a mix of user fees and subsidies. When requested, the Bank has also supported public-private partnerships in urban and rural water utilities, which have proved a valid option to turn around poorly performing water utilities and improve service quality and efficiency.[1] This approach fosters a virtuous circle whereby the utility improves its financial situation and gradually becomes able to finance a larger share of its investment needs. Experience shows that although concessions have worked in a few places, contractual arrangements that combine private operation with public financing of investment appear to be the most suitable option in many countries (Marin 2009). Finally, work is ongoing to assess the effectiveness of consumer subsidies in reaching and distributing resources to the poor; with evidence suggesting that connection subsidies may be a more effective way to target the poor than quantity-based subsidies (World Bank 2005b).

Water Charges in Irrigation. Evidence shows that irrigation demand is inelastic until prices rise to several multiples of the cost of providing the services. In practice, it has proven politically difficult to increase bulk water and irrigation prices sufficiently to move to the elastic part of the demand curve. As the most immediate demand management option, the Bank has thus favored the setting of water rights for surface and groundwater.

Prioritization of Water Lending. Management notes that although IBRD countries may borrow more for water in absolute terms given their country sizes, IDA countries receive more lending for water in relative terms. When looking at the level of water stress in IDA and IBRD countries, management notes that water-stressed countries receive proportionately more financing for water than non-water-stressed countries—in water-stressed countries, water constitutes 14 percent of total IBRD and IDA lending; in nonwater-stressed countries, water constitutes 9 percent of total IBRD and IDA lending.

Financing, Knowledge, and Reputation. Finally, management notes that project financing is only one way to address issues in the water sector. Several other mechanisms are used by the Bank to achieve its strategic vision of water for sustainable development. The Bank uses economic and sector work, policy dialogue, trust funds, and its reputation as an "honest broker" to engage client countries in complex water issues. For example, the World Bank Group has been engaged with McKinsey in looking at innovative tools to identify supply-side and demand-side measures that could constitute a more cost-effective approach to closing the water resource gap in countries and may even achieve budget savings in some components of the water sector. It is the combination of all these instruments that enables the Bank to provide assistance to countries that face the greatest water stress today and to address future water needs.

IEG Recommendations

Management welcomes and agrees with the IEG recommendations. These recommendations fit well with what the Bank is currently doing and can be accommodated within the framework of the existing water strategies, as the Water Resources Sector Strategy Mid-Cycle Implementation Progress Report will show.

IEG Recommendation	Management Response
1. Work with clients and partners to ensure that critical water issues are adequately addressed. • Seek ways to support those countries that face the greatest water stress. The mid-term strategy implementation review should suggest a way to package tailored measures to help the Bank and other donors work with these clients to address the most urgent needs, which will be far more challenging as water supply becomes increasingly constrained in arid areas. • Ensure that projects pay adequate attention to conserving groundwater and ensuring that the quantity extracted is sustainable. • Find effective ways to help countries address coastal management issues. • Help countries strengthen attention to sanitation.	**Ongoing/Agreed.** Management agrees with the recommendation, which is at the core of the *2003 Water Resources Sector Strategy*. The Bank has been responsive to government priorities on water in the most water-stressed countries and in those that will face problems in the future. Using a range of instruments (finance, knowledge, and reputation), the Bank has worked toward ensuring that its assistance adds value, especially vis-à-vis other development banks and donors. The Water Resources Sector Strategy Mid-Cycle Implementation Progress Report will highlight (i) how the World Bank has addressed client needs, differentiating by income group; (ii) the growing importance of addressing water issues at the river-basin level; (iii) areas of the 2003 strategy where the Bank has sequenced its approach, starting with studies, technical assistance, capacity building, and pilot projects to address complex issues, such as sustainable groundwater management and coastal management; and (iv) how the development community has been actively working toward meeting the sanitation MDG targets.
2. Strengthen the supply and use of data on water to better understand the linkages between water, economic development, and project achievement. • In project appraisal documents, routinely quantify the benefits of wastewater treatment, health improvements, and environmental restoration. • In project appraisal documents, routinely quantify the benefits of wastewater treatment, health improvements, and environmental restoration. • Support more frequent and more thorough water monitoring of all sorts in client countries, particularly the most vulnerable ones, and help ensure that countries treat monitoring data as a public good and make them broadly available. • In the design of WRM projects that support hydrological and meteorological monitoring systems, pay close attention to stakeholder participation, maintenance, and the appropriate choice of monitoring equipment and facilities. • Systematically analyze if environmental restoration will be essential for water-related objectives to be met in a particular setting.	**Ongoing/Agreed.** Management agrees with the recommendation, and the principle that more and better data would help to support efforts to improve the performance and accountability of the water sector, the results of Bank-financed water projects, and the impact of alternative water policies. Several global initiatives are under way (for example, IBNET, Hydrological Expert Facility), and efforts are ongoing as part of the standard evaluation analysis of projects to quantify the costs and benefits (and externalities). Better management and use of data will take place when the investment lending reforms are implemented. More specifically: • The Water Anchor will develop further core indicators for water projects (for example, sanitation/sewage, irrigation/drainage). • Regions will pilot new approaches to take advantage of new sources of information (such as remote sensing), tackling these with existing data sources. • Regions will scale up projects, building detailed information systems and benchmarking systems. • The Water Anchor and Water Sanitation Program will conduct an impact evaluation of sanitation and hygiene interventions at scale in achieving health and income outcomes. • As part of the development impact evaluation initiative, in collaboration with the Development Economics Department, the water sector will conduct further impact evaluations on health impacts of water and wastewater interventions. The Water Resources Sector Strategy Mid-Cycle Implementation Progress Report will outline how progress toward strengthening the supply and use of data will be monitored.

IEG Recommendation	Management Response
3. Monitor demand-management approaches to identify the aspects that are working or not working and to build on these lessons of experience going forward. • Clarify how to cover the cost of water service delivery in the absence of full cost recovery. To the extent that borrowers must cover the cost of water services out of general revenues, share the lessons of international experience with them so they can allocate partial costs most effectively. • Identify ways to more effectively use fees and tariffs to reduce water consumption. • Carefully monitor and evaluate the experience with quotas as a means to modulate agricultural water use.	**Ongoing/Agreed.** • Regions and the Water Anchor will examine financing of services delivery as part of Public Expenditure Reviews and other country-specific economic and sector work. • The Water Anchor and Regions will conduct a study on lessons learned about government payment for water services. • Regions will continue to explore fees, tariffs, and other options (metering, water rights, and the like) for demand management in Bank projects. • Regions will pilot evapotranspiration (ET)-based rights and community-based approaches to water resource management. A key priority of the Thematic Group on Water Resource Management, with the support of the Water Anchor, will be to document lessons learned on demand-management approaches.

On December 16, 2009, the Committee on Development Effectiveness (CODE) discussed the document entitled *Water and Development: World Bank Support, 1997–2007,* prepared by the Independent Evaluation Group (IEG), and *the Draft Management Response.*

Summary

The Committee welcomed the timely report, which will serve as an input to management's Water Resources Sector Strategy Mid-Cycle Implementation Progress Report (MCIPR). It also positively noted the constructive discussion between IEG and management on this evaluation. Members noted the continued relevance of the Bank's 1993 Water Resources Management Policy Paper and the 2003 Water Resources Sector Strategy. They commended the turnaround in portfolio performance in the water sector and encouraged continued efforts, particularly to further enhance project sustainability and focus on projects in Sub-Saharan Africa.

One area eliciting comments from almost all speakers was the sensitive and challenging issue of pricing and full cost recovery of water services. In this regard, different views were expressed on the roles of the public and private sectors in the provision of water, the potential for a market-based approach, and the importance of ensuring affordable water to the poor. Moreover, the discussions touched on issues of subsidies and transfers mechanisms, the coverage of operations and maintenance (O&M) costs, and the experience of Organisation for Economic Co-operation and Development countries in full cost recovery. In general, members supported the Bank's "pragmatic, but principled" approach to water pricing, outlined in the 2003 Water Resources Sector Strategy.

Members also intervened on specific topics such as the regulatory framework and institutions; the assistance to water-stressed countries/areas; the Bank's role in sanitation and groundwater management; links between water and climate change, and demography; and data collection and monitoring. The committee welcomed the comments from the regional staff who shared their country- and regional-level experiences. Some speakers expressed the view that the IEG evaluation could

have been a World Bank Group review of water assistance. IEG noted that IFC's investments in water have been small, only about 1 percent of IBRD/IDA lending. Others asked about aspects not covered in the evaluation report, such as the impact of the matrix management structure, the contribution of development policy lending (DPL), ongoing work related to innovative financing support, and key findings that were contained in the evaluation, but not explicitly referred to in the recommendations. Finally, comments and suggestions were made on the upcoming MCIPR of the Water Resources Sector Strategy.

Recommendations and Next Steps

Management is expected to present the Water Resources Sector Strategy MCIPR to the Committee in spring of 2010. Management confirmed that the MCIPR will cover the World Bank Group water sector activities, even though the 2003 Water Resources Sector Strategy covered IBRD and IDA only. This IEG evaluation report and Management Response will be publicly disclosed.

Main Issues Discussed

Pricing and Cost Recovery. The Committee noted the complexities of water pricing and the political sensitivities around full cost recovery. In general, members supported the Bank's "principled, but pragmatic" approach in water pricing, though several speakers, mindful of the difficulties of a full market-based approach, indicated that the existence of price-cost gaps requires it to be fully quantified jointly with the fiscal resources needed to ensure full cost recovery. They took note of management's clarifications on the definition of cost recovery and the difficulty of achieving full cost recovery even in high-income countries that have the political will to embrace this goal. Some members commented that this matter needs to be considered as part of a comprehensive and coherent set of interventions, including consideration of social issues. IEG pointed out that less than one-fifth of water supply and sanitation projects that set out to recover costs either partially or fully have succeeded, and that it is important to clarify how to cover the costs of water service delivery in the absence of full cost recovery. Noting the importance of ensuring affordable access to water, especially for the poor, a member cautioned about pursuing a market-based approach to water supply and sanitation and the continued important role of government. Some others, however, commented on the potential of the private sector to provide efficient services, if accompanied by appropriate legal frameworks and regulatory institutions. Some specific comments were made about how to cover O&M costs in the absence of sufficient revenue collection and fiscal space, and on the use of subsidies or direct transfer to provide affordable access to water for the poor, on the relatively low number of projects where economic rate of return is considered, and the potential of improving cost recovery by reducing unaccounted-for water usage.

Data. There was general agreement on the importance of data for project design and monitoring and evaluation in light of IEG's findings that few projects that plan to collect data actually follow through. One member raised a concern regarding balancing data sharing with national security concerns, emphasizing that data provision should not be a condition for a project. In this respect, the significant effort by the international water community to collect data, which could be utilized, was noted. Stressing the importance of measuring the outcomes, a suggestion was made to benchmark the impact of the Bank's operations against the Millennium Development Goal and an indicator that would measure the efficiency in use of water resources.

Water-Stressed Countries. While supportive of IEG's findings on the need to adequately address critical water issues in water-stressed countries, a few members raised the issue of how these countries are defined and the complexity of this subject. IEG clarified that the Water Poverty Index was based on five components—resources, access, capacity, use, and environment. Others encouraged management to attend to the water supply and sanitation needs of the poor, while integrating institutional, financial, and social dimensions into the Bank's interventions.

Mid-Cycle Implementation Progress Report. Members made a number of requests for consideration in the MCIPR, including the Bank's role in strengthening institutions, knowledge sharing, use of DPLs, safeguards issues, matrix management structure, and use of incentives. Management was encouraged to keep in mind the impact of climate change on water resources and their use, as well as the nexus between demography and water. Management said that it would review the World Bank Group operations, in response to interest expressed for a broader coverage to include private sector issues. Management briefly commented on the Bank's interventions in the water sector encompassing not only investment lending but also DPLs and analytical work. Management spoke on the new Water Sector Board and the highly cross-sectoral nature of its intervention in the water sector.

Giovanni Majnoni, Chairperson

The members of the External Advisory Panel acknowledge the colossal amount of work achieved by the Independent Evaluation Group team in preparing this report on the World Bank's support to water and development. Thirty-one percent of all Bank projects approved since fiscal 1997 have at least one water activity, so that the portfolio examined by this evaluation consists of 1,864 projects financed between fiscal 1997 and the end of calendar 2007, for a total of over $118 billion. That the evaluation team took on this range of activities and examined it through multiple lenses—25 separate Issues Papers were researched and written for this study—is a testament to their seriousness and thoroughness. The Advisory Panel appreciates the care with which this team sorted, analyzed, and reanalyzed the data gathered, and we urge Bank management to apply similar rigor in addressing the recommendations that come from this impressive work.

The basic message of this report is a message of urgency. Water is a limited resource on which life and development depend. As climate changes and population grows, business as usual is not an option. The social, economic, and political consequences of water shortages, exacerbated by distributional unevenness, are real, as are the effects of water-related hazards. Few institutions are as well positioned as the World Bank to assemble the resources and the partnerships required to address such global concerns.

It is important to recognize that World Bank water projects over all have recently shown improved success rates. However, there is no room for complacency, because, as the evaluation points out, much of the work has been concentrated on regions and on problems that are somewhat easier to deal with. As the report says, "The Bank and the countries have not yet sufficiently tackled several tough but vital issues, among them sanitation, fighting pollution, restoring degraded aquatic environments, monitoring and data collection, and cost recovery."

One particularly difficult problem concerns the sustainability of investments over time; the use of inappropriate technologies, failure to secure institutional changes and establish clear postproject accountabilities, and lack of attention paid to cost recovery are just some of the reasons why sustainability may not be achieved. To ensure that investments yield long-term benefits in terms of economic and social development, it may well be that a project focus will

not suffice. Instead the Advisory Panel urges Bank management to develop a longer-term, more comprehensive, multisectoral and process-focused approach to its water support. The evaluation provides numerous examples from the Bank's own experience showing not only that such an approach is doable, but that its results are more lasting and significant for more people than those of separate projects with shorter timelines. The Panel would also wish the Bank to take a more proactive approach to its lending and to assess whether some of the current lending trends will be appropriate in the future, given changing socioeconomic and environmental conditions; for example, the dwindling number of project approvals for coastal management work could be reviewed in the context of demographic and potential climate change.

As the report says, the ultimate goal is to balance the resources available with societal requirements. We underline this statement by noting that this goal is not just for some countries, but for all countries, and not just for some people, but for all people. This goal requires a long-term, strategic engagement of Bank management and Bank staff with counterparts in other agencies and around the world to address usage and wastage, retention and pollution, balancing today's needs with those of future generations.

A careful reading of the findings of this thorough report provides strong guidance for Bank management on how to develop and pursue such a strategic engagement.

Members of the Advisory Panel especially urge attention to the following findings and opportunities:

1. Integrated water resources management is an ongoing process, not an ad hoc program, and as such deserves constant reinforcement through technical and political requirements. Working with borrowers, Bank staff must assiduously advise and encourage an integrated, multisectoral analysis of water problems and the development of water solutions that recognize the relationships between water and the economic development and poverty reduction objectives of governments.

2. In spite of its comparative advantage, the Bank appears to have reduced the emphasis given to the economic analysis of water projects. Data collection is inadequate (especially as it deals with the social and developmental impacts of water efforts), monitoring is inconsistent,

and cost recovery systems are too often assumed rather than rigorously pursued. Concerns over the sustainability of projects, but even more important over the sustainability of access to water, require that such analysis be reinstated and followed closely.

3. Having urged more attention to cost recovery and the financial sustainability of projects, the Advisory Panel nonetheless urges the Bank to develop strategies that will bring poorer countries that face greater water crises into the lending portfolio. More-innovative funding packages, developed in conjunction with partners, need to be considered for the poorest countries facing the greatest water stress. As noted above, the unevenness of the distribution of water surpluses and shortages has both economic consequences and political and social consequences. These cannot be ignored. A fuller focus on the developmental impacts of water—its shortages or its accessibility—would suggest greater attention to integrating water programs with other sectors in countries with the greatest water stress and the fewest resources for addressing this stress.

This IEG report has come at a critical time, and we are grateful for the opportunity to have worked with the IEG evaluation team in considering the implications of its findings. The report provides a strong, well-researched basis for the Bank's development of a more integrated, far-looking approach to one of the world's most urgent challenges.

Signed:

Mohamed Ait Kadi
Chair of the Global Water Program Technical Committee
President of the General Council of Agricultural Development,
Ministry of Agriculture, Rural Development and Fisheries, Morocco

Mary B. Anderson
Executive Director of CDA Collaborative Learning Projects

Judith Rees
Professor of Environmental and Resources Management
Director of the Grantham Research Institute on Climate Change and the Environment
London School of Economics

Peter Rogers
Gordon McKay Professor of Environmental Engineering in the School of Engineering and
Applied Sciences at Harvard University
Senior Advisor to the Global Water Partnership
Fellow of the American Association for the Advancement of Science
Member of the Third World Academy of Sciences

Chapter 1

EVALUATION HIGHLIGHTS

- The supply of water is unevenly distributed and subject to new demands.

- Water shortages contribute to conflicts, and the lack of clean water and sanitation has huge health costs.

- In the 1980s the Bank focused on water services infrastructure, and in the 1990s on improving management; since 2001 the Millennium Development Goals have been a focal point.

- With a view to contributing to greater effectiveness, the evaluation covers water resources management, water and environmental sustainability, and water service delivery, as well as the institutions that coordinate work within and among those areas.

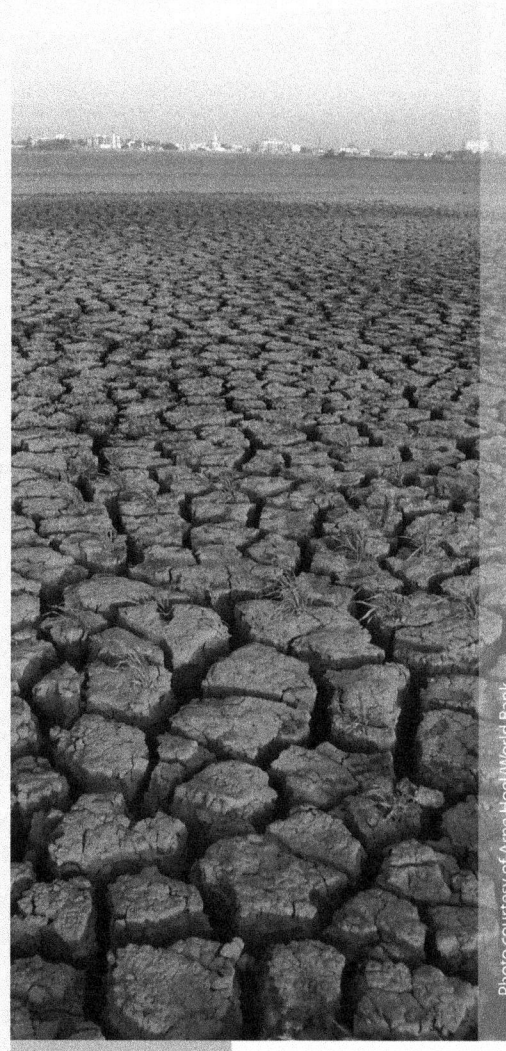

Photo courtesy of Arne Hoel/World Bank.

Water and the World Bank

Global Context

The amount of water in the world's ecosystem has been roughly constant for millennia, but over that time the planet has added 6 billion people. Moreover, the usable supply of water is limited: only 3 percent of the world's supply is freshwater, and two-thirds of that is locked in glacier ice or buried in deep underground aquifers, leaving just 1 percent readily available for human use.

What water is available is unevenly distributed. Water shortages are an ever-present threat in the world's more arid regions. About 700 million people in 43 countries face water stress, unable to obtain the 1,700 cubic meters of water per person per year that is considered a minimum supply (UNDP 2006, p. 14), based on estimates of water requirements in the household, agricultural, industrial, and energy sectors and the needs of the environment.

Development patterns, increasing population pressure, and the demand for better livelihoods in many parts of the globe all contribute to a steadily deepening global water crisis that will severely constrain future economic progress.[1] Major rivers such as the Nile, the Jordan, the Yangtze, and the Ganges are overtaxed and regularly shrink for long periods during the year. Underground aquifers below New Delhi, Beijing, and many other quickly growing population centers are falling rapidly. Other water basins in trouble include the Balsas and the Grande de Santiago in Mexico, the Limpopo in southern Africa, the Hai Ho and the Hong in China, the Chao Phraya in Southeast Asia, and the Brahmani and eight other rivers in India (Revenga and others 2000).

Water crises can also lead to sudden massive destruction. Over the period 1973–2006, about 3.5 billion people were affected by floods and heavy rains from tropical storms,[2] disasters that are increasing in number at an annual average rate of at least 5 percent. To this must be added the increasing likelihood of climate changes, which are already disrupting rainfall patterns, feeding ever more powerful windstorms, and creating droughts of unprecedented severity and frequency. There is scientific consensus that, under business as usual, these challenges will worsen dramatically (Gleditsch, Nordàs, and Salehyan 2007, p. 2).

Water is limited in supply, unevenly distributed, and under growing pressure.

Extended droughts already have hit the southeastern United States, northern China, Ukraine, Europe's Mediterranean belt, the Jordan River region, the Republic of Yemen, parts of the Sub-Sahara, and Australia's Murray-Darling Basin—the source of a third of that country's food supply. Drought is pushing food prices higher in many countries and has caused some to hoard farm exports. Six hundred million subsistence farmers have no access to irrigation water and remain trapped in poverty, according to *World Development Report 2007* (World Bank 2007b).

Competition for and contamination of water resources have local and regional political impacts and create the potential for conflict. For example, drought has pushed competing herders and farmers into conflict in Chad and Sudan, contributing to the misery in Sudan's war-ravaged Darfur region (Borger 2007).

Competition for water contributes to conflicts.

A host of environmental issues affect water resource availability: deforestation, watershed degradation, encroachment on recharge areas, pollution from point and nonpoint sources, infestation by aquatic weeds, inadequate environmental flows, and drought and floods, among others. To some officials and commentators, the water crisis is about absolute shortages of supply. But the United Nations' *Human Development Report 2006* (UNDP 2006) rejected that view, arguing instead "that the roots of the crisis in water can be traced to poverty, inequality and unequal power relationships, as well as flawed water management policies that exacerbate scarcity" (UNDP 2006, p. v).

Hundreds of millions of the world's poorest do not share even minimal access to safe drinking water and basic sanitation services. Each day millions of women and young girls collect water for their families—a ritual that reinforces gender inequalities in employment and education, the UN *Human Development Report* noted (UNDP 2006, p. v).

The health impacts of the pervasive lack of water supply and sanitation services are staggering. Each year almost 2 million children in the developing world die from lack of clean water and little or no sanitation. About one-third of people who lack access to drinking water live on less than $2 a day, the United Nations reports. Some 385 million people must try to survive on $1 a day or less. There is little hope that these households can pay for water connections (UNDP 2006, p. 48).

The health impacts of the lack of water supply and sanitation are staggering.

Water Milestones and the Bank's Commitments

The 1992 Rio Earth Summit—and the International Conference on Water and the Environment, where the Dublin Principles[3] were adopted earlier that same year—raised awareness of growing water scarcity. Then, in 2000, at the Second World Water Forum in The Hague, the Netherlands, the Global Water Partnership called for stronger international dialogue, better capacity building, and additional financial resources. That same year the international community adopted the Millennium Development Goals (MDGs) at the United Nations Millennium Conference.[4] The MDGs for poverty, hunger, and health relate indirectly to water, but the environmental goal addresses water directly, seeking to halve the number of people without access to safe drinking water and basic sanitation by 2015.

In response to the MDGs, and following up on the Second World Water Forum, the World Water Council and the Global Water Partnership issued a report for the Third World Water Forum in 2003 in Japan. The report, *Financing Water for All: Report of the World Panel on Financing Water Infrastructure* (World Panel on Financing Water Infrastructure 2003), stressed that attaining the MDGs would require a doubling of the resources allocated to the task, as well as improvement of governance, better cost recovery, and some national public funding.

More recently, the United Nations has called upon the development community to respond to the myriad challenges posed by misuse of water and its resulting inequalities. The years 2005–15 were designated the Water for Life Decade by the United Nations. In line with this, during the 2007 Spring Meetings of the World Bank Group and the International Monetary Fund, the World Bank Group committed itself to partnering with the United Nations Development Programme (UNDP), DFID (the United Kingdom's Department for International Development), and other donors to confront the crisis in water supply and sanitation.

Evolution of the World Bank's Strategic Approach to Water

In the 1980s the World Bank focused on water services infrastructure development as part of its core business. The decade saw large gains in the number of families served by a safe water supply (mostly in Asia), although little progress was made in sanitation. Irrigated areas were expanded, and dams helped to offset the results of climate variability and contributed to energy supplies (Parker and Skytta 2000, p. 3). However, the exclusive focus on infrastructure posed serious environmental, social, and financial sustainability issues.

During the 1990s the Bank's focus shifted toward improving the management of water utilities, irrigation, rural water systems, water resources, and land use. The Bank's 1993 Water Resources Management Policy Paper (evaluated by IEG in 2002) moved the institution away from infrastructure development. The paper also shifted the Bank's planning process from one based on discrete investments within the sector to a multisectoral approach. The paper focused human and financial resources on three complementary roles: designing projects that would develop a stock of infrastructure for multiple water uses; establishing or strengthening river and lake basin management institutions; and helping to craft policies for the rational management and use of transboundary water.

The 1993 Water Resources Management Policy Paper moved the Bank toward a multisectoral approach to water planning.

With the turn of the century, the Bank's approach again shifted, to one of balancing infrastructure and management-focused investments. In 2001 the World Bank Group committed itself to achieving the MDGs. With the 2001 Environment Strategy, the 2003 Water Resources Sector Strategy, the 2003 Infrastructure Action Plan, and the 2003 Water Supply and Sanitation Business Strategy, water was given more prominence. The 2003 Water Resources Sector Strategy focused on putting the 1993 principles into practice, emphasizing the importance of infrastructure finance.[5] A main message of the 2003 strategy was that the Bank needed to continue its efforts toward integrated water resources management (IWRM; see box 1.1). The strategy was also the first to highlight the impact of climate change on the sector.

> After 2001 the Bank's focus shifted to meeting the MDG targets. Its 2003 sector strategy called for continued efforts toward IWRM.

The 1993 and 2003 sector strategy papers are complementary, and together with the Bank's poverty reduction mandate they have helped inform issues of supply and improve the performance of utilities and user associations. Appendix table C.1 compares the main provisions of these strategic documents.

Scope and Purpose of the Evaluation

Water has always been central to World Bank Group work, and the International Development Association (IDA) and the International Bank for Reconstruction and Development (IBRD; the two are jointly referred to here as the Bank) in particular have supported countries in many water-related activities. This evaluation covers water resources management and development, water-related environmental sustainability, and water services delivery. Water resources management includes the management of the demand for water, environmental sustainability includes managing en-

vironmental flows and water quality as global public goods, and water services delivery includes irrigation and drainage, water supply and sanitation, and hydropower. Interactions within and among the three areas are influenced by the institutions that coordinate development.

The evaluation covers so broad an area because most actors in the development community have transformed and broadened their approach to water. As stresses on the quality and availability of water have increased, the donor community has moved steadily toward considering each water-related activity in light of other competing uses.

The evaluation analyzes the full universe of Bank-financed or Bank-administered water activities approved or closed from July 1, 1996 (the beginning of the Bank's fiscal 1997), to January 1, 2008. IEG identified 1,864 projects approved or completed over the 11.5-year period (including those approved or completed by the Global Environment Facility, GEF) that include at least one water-related activity. For each water-related issue evaluated, all relevant projects were included in the analysis.

> The study analyzed the full universe of Bank-supported water activities from fiscal 1997 through the end of calendar 2007, a total of 1,864 projects.

Given the scope of the topic, the evaluation is also a meta-evaluation. It draws on recent IEG evaluations on water-related themes as well as relevant findings from recent studies by the Bank's development partners. It also draws on sectoral and thematic evaluations and impact studies carried out by bilateral and multilateral donors, nongovernmental organizations, and others, as well as work conducted by Bank Operations as part of their self-evaluation (see appendix B on the evaluation methodology and the Bibliography).

BOX 1.1

BANK STRATEGY AND INTEGRATED WATER RESOURCES MANAGEMENT

IWRM calls for the integration of actions affecting drinking water and sanitation supply, agriculture, irrigation, hydropower and other energy production, and maintenance of environmental water flows to protect habitats and sustain groundwater supplies. Under IWRM, the results of water management programs are monitored to permit ongoing adjustments to strategies and practices.

IWRM leads toward the recognition that water policy is bound up together with government policies on security, economic development, food production, public health, and other essential governance missions. It is best conceived of as a framework to guide thinking about the management of water resources so that the approach taken varies appropriately according to geography, climate, and institutions. It is inherently an approach that countries themselves should be encouraged to take—not one that applies only to the Bank's own work.

Source: World Bank (2003a).

IEG carried out more than 30 background and case studies of 6 countries and 1 water basin study (these are listed in appendix B). These efforts analyzed Bank activities by thematic area and by type of activity. All the papers looked at the same universe of projects. In thematic areas where there is little strategic guidance, the evaluation analyzed what was actually being done, so as to distill the institution's revealed and evolving preferences. The evaluation is by definition retrospective, but it identifies changes that will be necessary in the future, including those related to institutional and financial sustainability.

Since the commitment to water resources management and water services is largely a concern for IDA and IBRD, they were the focus of the evaluation. The evaluation did not conduct a comprehensive review of investments by the International Finance Corporation, also a member of the World Bank Group, in this area.

The evaluation addressed basic questions: What is the Bank doing in the water sector? Where is it doing those things? How has the Bank's approach changed over time (and in light of strategy documents and other commitments)? What activities have tended to achieve their objectives most consistently?

Structure of This Report

This report provides only a broad overview of the substantial research generated by the evaluation. Organizing this wealth of material presents considerable challenges, as there are many disparate parts, some of which may seem only loosely related to all the other parts.

Any framework for organizing this material is necessarily artificial and excludes some important work. The report's structure hews closely to the three large areas identified earlier—water resources management, environmental sustainability, and water services delivery—which are connected through the institutional context that is necessary to coordinate work in all three of those areas.

The report begins by establishing the characteristics of the full set of Bank lending activities in water and its performance (chapter 2). It then more closely examines the Bank's work related to water resources management (chapter 3), environmental sustainability (chapter 4), water services delivery (chapter 5), and institutions and coordination (chapter 6). The report concludes with a summary of the findings and with conclusions and recommendations intended to inform the Bank's ongoing Mid-Cycle Implementation Progress Report and future updates of the Bank's water strategy.

Chapter 2

EVALUATION HIGHLIGHTS

- Bank lending for water has increased since 1997 by over 50 percent.

- Over the 11-year evaluation period, 31 percent of Bank projects have been related to water.

- Bank-financed water work is concentrated in a subset of countries, although the number of borrowing countries has grown steadily over time.

- Performance of the water portfolio has improved considerably in the last five years.

- Of the 550 projects dealing with water supply and sanitation, a little more than half were implemented by sector boards other than that for water and sanitation, indicating considerable integration.

- During the period studied, projects implemented by the Water Supply and Sanitation Sector Board improved from 63 percent moderately satisfactory or higher to 89 percent.

The Bank's Water-Related Activities and Their Performance

A large part of what the Bank finances has something to do with water. Between the Bank's fiscal year 1997 and the end of calendar 2007, the Bank approved or completed 1,864 projects and grants that included at least one water-related activity.

Total funding for these projects and grants was $118.4 billion, of which direct Bank support to water was about $54.3 billion. The average loan (exclusive of grants and other non-lending support) was $67 million. As might be expected, direct support is so much smaller than total lending and grants because, although the portfolio includes many projects completely focused on water-related activities, it also includes projects in which those activities were only subcomponents of a larger project whose main focus was not on water. Overall, 424 distinct water-related activities had defined and measurable results (see appendix D). But the scale of these activities varies, and some of the projects are among the largest infrastructure loans to the Bank's largest borrowers.

Overview of the Water Portfolio

Lending for water grew 55 percent in commitments during the period from fiscal 1997 to the end of calendar 2007. Lending for most water subsectors dipped to a low point in fiscal 2002, but the rhythm of project approvals subsequently picked up as part of the overall increase in Bank infrastructure lending (figure 2.1).

Bank lending for water has increased since 1997: 1,864 projects worth $118.4 billion have been approved or completed.

From fiscal 1997 to end-2007, 1,317 water-related projects were *approved*, representing a total Bank commitment of $70 billion. For comparison, the Bank approved a total of 4,204 projects and grants worth $248 billion during the same period. Water-related projects thus represented 31 percent of all Bank projects approved and 28 percent of all Bank funding commitments over the evaluated period. The Bank was also involved in 238 GEF water projects, all but 34 of them associated with a Bank loan or credit. In addition, 40 projects were carbon offset, Montreal Protocol, or rainforest projects.[1] Between 1984 and 1996, 547 water projects were approved that closed *after 1997*. Since 1997, 960 water projects and grants have been *completed* (including those approved before 1997).[2]

Since 1997, 31 percent of Bank projects and 28 percent of all lending commitments have been related to water, but the average project has been getting smaller.

Although the Bank today funds more water projects than ever before, the average project is somewhat smaller than in the late 1990s, in keeping with the overall Bank trend away from larger loans and toward less variability in lending (figure 2.2).[3]

IEG reviewed the objectives of all projects and identified those that had half or more of their objectives focused on water. Six hundred and sixty-two projects, about 35 percent of the 1,864 projects in the portfolio, fit this description and will be referred to as *dedicated projects* in this evaluation (table 2.1).

About 35 percent of the water-related projects had at least half their objectives devoted to water.

Table 2.1 also shows the number and commitment amount of projects overseen by the Water Supply and Sanitation (WSS) Sector Board (renamed the Water Sector Board in 2007; see appendix D). The majority of the dedicated projects (56 percent of the 662 projects) were under the purview of sector boards other than WSS. Examples include irrigation projects done under the Rural Sector Board, lake restoration projects done under the Environment Sector Board, and flood emergency projects done under the Urban Sector Board.

This pattern held for the portfolio as a whole and for non-lending services. The WSS Sector Board was responsible for fewer than half of the water-related projects that had fewer than half of their objectives focused on water. A recent review of Bank analytical and advisory activities from fiscal 2003 to 2009 done during preparation of the Water Mid-Cycle Implementation Progress Report found that analytical work is being mainstreamed effectively: only 84 of 606 pieces of water-related analytical and advisory activity funded in the Bank's budget were completed by staff mapped to the water sector.

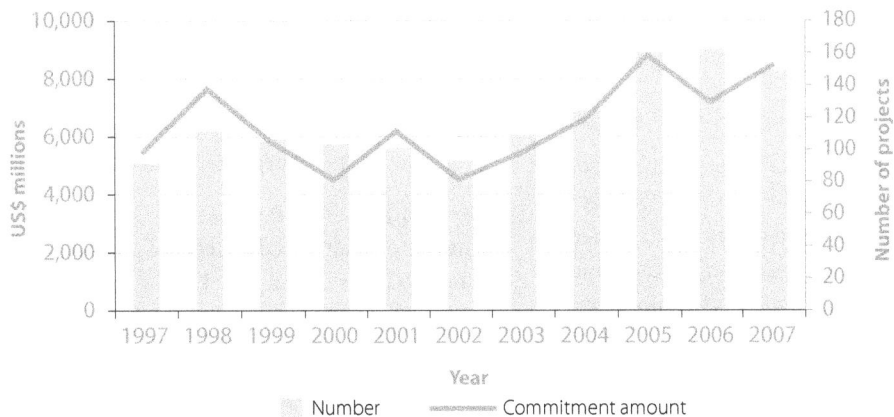

Source: IEG water database.

Note: Total commitment amounts given for all projects. Years are fiscal years. The total number of projects approved was 1,317; the 1,864 projects covered in the evaluation include projects that closed during the evaluation period but were approved earlier.

Trends in the Water Portfolio

Lending by Sector

Because water is a crosscutting theme and the integration of water was a key element of the Bank's water strategy, the degree to which it has appeared in the work managed by sector boards other than the WSS Sector Board is impor-

tant (figure 2.3). The Rural Sector Board, responsible for the largest share of Bank lending, has overseen the most water-related projects: 24 percent of the portfolio. Many agricultural projects have routinely included irrigation, drainage, or flood mitigation alongside other water-related activities such as watershed management, forestry, and drought miti-

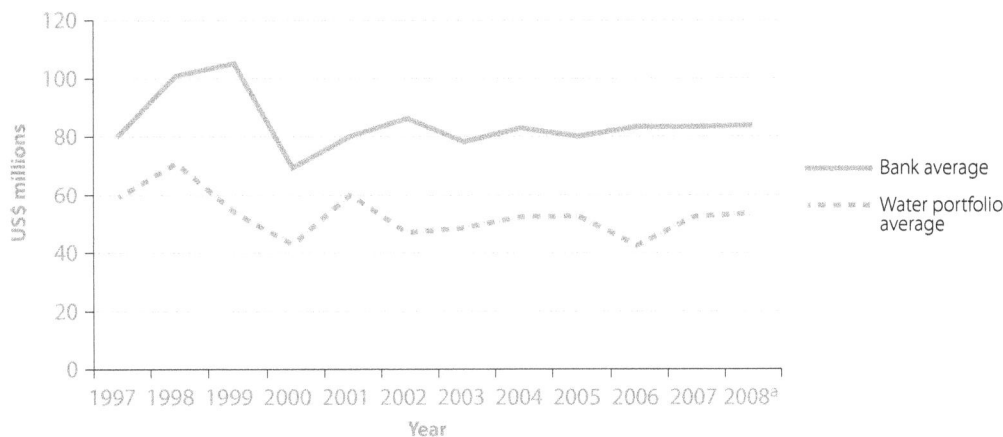

Source: IEG water database.

Note: Includes only loans (including Development Policy Loans) approved during the period.

a. Includes small number of 2008 water projects.

TABLE 2.1	The Water Portfolio and Its Dedicated Projects	
	Number of projects	Commitments (US$ millions)
Entire portfolio	1,864	118,420
Overseen by WSS Sector Board[a]	293	20,615
Projects with mainly water objectives (dedicated projects)	662	41,214

Source: IEG water database.

a. Appendix figures J.1–J.3 provide an analysis of ratings for the WSS Sector Board–supervised portfolio.

gation. Other sector boards that have been highly involved are Environment (18 percent of the portfolio), Energy and Mining (12 percent), and Urban Development (11 percent). This analysis suggests that the integration of water practice envisioned in the Bank's water strategy is well under way.

The distribution of water projects among the various sector boards suggests that the integration of water practice is well under way.

Lending by Focal Area

The Bank works with its borrowers to achieve many different water objectives, and it follows a variety of approaches to meet those objectives. Table 2.2 lists, within three broad subsets of the water portfolio, the focal areas with the largest number of projects. The list gives an idea of the breakdown of the overall portfolio. Some of these focal areas have been

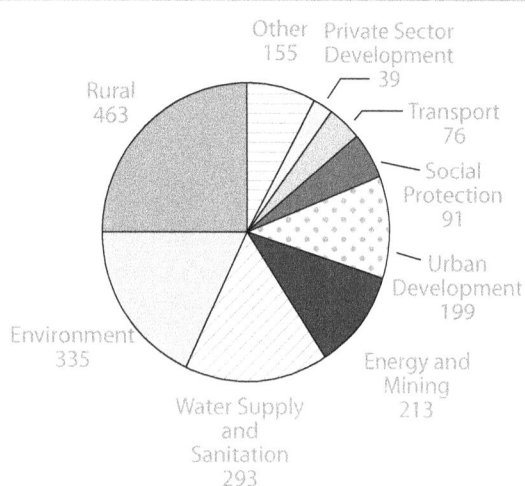

| Figure 2.3 | Projects with Water-Related Activities by Implementing Sector Board |

Other 155
Private Sector Development 39
Rural 463
Transport 76
Social Protection 91
Urban Development 199
Environment 335
Energy and Mining 213
Water Supply and Sanitation 293

Source: IEG water database.

the subject of in-depth research for this evaluation; for others the counts were compiled using Bank database codes or word searches. Many of the focal areas overlap, leading to double counting, but each provides a filter through which to look at water issues.

The largest subsets of projects are in irrigation, wastewater treatment, and dams and hydropower.

The two largest focal areas by number of projects were wastewater treatment and irrigation. However, the largest amounts of money went to projects involving irrigation and hydropower or dam activities. (Because dams can support irrigation, the two categories are not mutually exclusive.)[4]

Lending by Region

Analysis of lending and patterns of practice by Bank Region shows that the Africa and Latin America Regions led in numbers of projects approved, with 456 and 407 projects, respectively (figure 2.4).

However, because projects tended to be larger in East Asia, that Region led in aggregate borrowing. Africa ranked fourth in amount of money committed, indicating that projects in Africa tended to be small, a reflection of the limited IDA lending envelope for individual African countries and the countries' limited absorptive capacity.

Attention to the various focal areas listed in table 2.2 varies considerably by Region, as detailed in appendix J.

Concentrations in Lending Patterns

A total of 142 countries turned to the Bank for water lending during the evaluation period; the top 10 accounted for 579 projects (31 percent of the total; see appendix table J.1 for a ranking by number of projects). Although the number of countries that borrow for water has varied from year to year, the trend has been upward: in 1997 a total of 47 countries were served, but by 2007 there were 79 borrowers. Large IBRD or blend (IBRD/IDA) borrowers made up most of the list; among countries eligible to borrow only from IDA, only Vietnam was among the top 10 borrowers for water (table 2.3).[5]

The top 10 borrowers for water accounted for 31 percent of projects and 56 percent of commitments.

The concentration of Bank-financed water work in a few countries was even more apparent in commitments: 56 percent of all Bank commitments for water went to the top 10 borrowers (table 2.3). To a large degree, this pattern mirrored Bank lending generally: Bank-wide, the top 10 borrowers (not the same 10 countries as for water) accounted

TABLE 2.1 Water Projects by Focal Area

Subset and focal area	Number of projects[a]	Total commitment[b] (US$ millions)
Water and land		
Irrigation	311	26,490
Groundwater[c]	229	20,508
Hydropower or dams	211	21,800
Floods	177	15,509
Droughts	110	9,960
Water supply and sanitation		
Urban water supply	229	15,522
Rural water supply	218	13,871
Wastewater treatment	312	13,460
Urban sanitation and sewerage	190	15,609
Rural sanitation and sewerage	108	5,894
Environment		
Watershed management	218	13,100
Rivers and lakes	174	14,780
Coastal zones	121	4,660
Inland waterways and ports	104	7,632
Fisheries	87	5,034

Source: IEG water database.

a. Includes both dedicated and nondedicated projects.

b. Total commitments refer to loan amounts for all projects that include the indicated activity.

c. Includes 89 projects focusing on aquifer conservation or protection.

FIGURE 2.4 Water Projects and Commitments by Region, 1997–2007

Source: IEG water database.

Note: AFR = Sub-Saharan Africa, EAP = East Asia and Pacific, ECA = Europe and Central Asia, LAC = Latin American and the Caribbean, MNA = Middle East and North Africa, SAR = South Asia.

for about 52 percent of commitments. The largest borrower for water, in terms of both number of projects and commitments, was China, with 133 projects and almost $19 billion—16 percent of total commitments—in the evaluated portfolio. Bank lending to China for all sectors was somewhat less concentrated, at about 7 percent of the total. Brazil was second in number of water projects but third in commitments, where it was preceded by India.

An Indian irrigation project was the largest of the water-dedicated projects in the portfolio and one of two Indian projects was in the top 10 (table 2.4); China had 5 projects among the top 10. All of the top 10 projects were IBRD loans (table 2.5 lists the top 10 water projects funded by IDA), and all but two were approved in the 1990s, confirming the observed recent trend toward smaller projects. Hydropower projects dominate the list, even though the Bank was not funding hydropower in the early part of this decade.

The largest projects tended to be in hydropower, although the Bank was not prioritizing that sector in the early part of the 2000s.

Portfolio Performance: Ratings against Objectives

Over the evaluation period, of the set of 857 completed projects (dedicated and nondedicated), 77 percent had an aggregate IEG outcome rating of moderately satisfactory or higher (hereafter referred to as *satisfactory*) when measured against stated objectives.[6] This was just above the Bank-wide average of 75 percent satisfactory during the same period.[7] When only dedicated water projects are considered, the aggregate IEG outcome rating over the evaluation period was 75 percent satisfactory.

The water portfolio as a whole is performing on par with other Bank projects.

In the aggregate, ratings by focal area indicate that the best performers have been projects for waterways and ports and for flood reconstruction (appendix figure J.6). Urban water supply, sewerage, and sanitation projects performed worse on average than other water-related projects. This may reflect the generally below-average performance of the water sector during fiscal 1997 to end-2002. At 66 percent satisfactory (11 percentage points below the sector average), urban dedicated water projects were rated satisfactory less often than all others. Except for those in sanitation, projects with water-related activities in rural environments generally performed better than those in urban areas.

Urban water supply, sewerage, and sanitation projects tended to perform worse than other water-related projects.

When the evaluation period is divided into two subperiods, 1997–2002 and 2002–07, the picture changes considerably. The outcome ratings of almost all types of water projects improved in the second five years over the first. Figure 2.5 compares the ratings of water-dedicated projects by focal area in the two subperiods. The most improved areas were rural sanitation and sewerage and rural water supply, which also had the highest ratings in the more recent period. The only focal area exhibiting a decrease in the percent of satisfactory outcomes was inland waterways and ports. Also among those receiving the highest recent ratings were projects involving floods, groundwater, and coastal zones.

In the aggregate, ratings by Bank Region (figure 2.6) indicate that the poorest-performing Region was Africa, where the

TABLE 2.3	The Top 10 Borrowers for Water		
Country	Total amount in water portfolio (US$ millions)	Water portfolio ranking	Entire Bank portfolio ranking
China	18,840	1	2
India	13,993	2	1
Brazil	7,153	3	3
Indonesia	5,637	4	7
Mexico	5,556	5	6
Pakistan	3,931	6	10
Argentina	3,199	7	5
Russian Federation	2,984	8	8
Vietnam	2,740	9	11
Turkey	2,211	10	4

Source: IEG water database.

TABLE 2.4	The 10 Largest Water Projects		
Country	Total amount (US$ millions)	Project name	Approval year
India	485	Tamil Nadu Irrigated Agriculture Modernization and Water Bodies Restoration and Management Project	2007
China	460	Xiaolangdi Multipurpose	1994
China	430	Xiaolangdi Multipurpose II	1997
China	400	Ertan II Hydroelectric Project	1996
China	400	Wanjiazhai Water	1997
Mexico	400	Irrigation Sector	1992
India	394	Madhya Pradesh Water Sector Restructuring	2005
China	380	Ertan Hydroelectric Project	1992
Turkey	369	Emergency Flood Recovery	1999
Mexico	350	Second Water Supply & Sanitation Project	1994

Source: IEG water database.

Bank financed 456 water-related projects (appendix figure J.7). Of these, 194 are closed, and outcomes for 62 (32 percent) were rated unsatisfactory. Africa also had the largest number of highly unsatisfactory projects (6), although the Middle East and North Africa, with 5 highly unsatisfactory projects out of a smaller set of water projects, performed worse in percentage terms. (Appendix E provides more information on the analysis of highly satisfactory and highly unsatisfactory projects.) Africa was also the worst-performing Region in sustainability ratings, with only 56 percent of completed projects classed as resilient to foreseeable risks.

Performance of the water portfolio improved substantially in the last five years of the period. Africa posted the strongest improvement.

When broken into subperiods, however, performance shows a clear improvement in the last five years in all Re-

gions except Europe and Central Asia, although the difference was not always statistically significant. Performance in Africa improved by 23 percentage points (figure 2.6).

Integrating Water

There are two basic ways in which the Bank's water-related activities can be integrated. One is thematic integration within the water sector itself. The other is integration of water into the work of other sectors. Both are of concern for IWRM.

In reviewing the integration of water into the projects of other sectors, the evaluation gave special attention to the more traditional water activities: water supply, sanitation, and sewerage (WSS). As noted earlier, of the 550 projects dealing with WSS, 273 were supervised by the WSS Sector Board and 277 by other sector boards. An updated review of all projects approved through fiscal 2009 shows that this trend is continuing, with fewer than a fourth of all active

TABLE 2.5	The 10 Largest IDA Water Projects		
Country	Total amount (US$ millions)	Project name	Approval year
Pakistan	285	National Drainage Program	1998
India	283	Tamil Nadu Water Resources Consolidation	1995
China	200	Forest Resource Development	1994
Bangladesh	200	Emergency Flood Recovery Project	1999
Tanzania	200	TZ-Water Sector Support SIL	2007
Nigeria	200	Second National Urban Water Sector Reform	2006
China	188	Irrigation Agriculture Intensification	1991
India	181	Mahar Rural Water Supply and Sanitation	2004
India	160	Upper Krishna Phase	1989
India	160	Maharashtra Irrigation I	1986

Source: IEG water database.

FIGURE 2.5 Outcomes of Water Projects by Focal Area and Subperiod

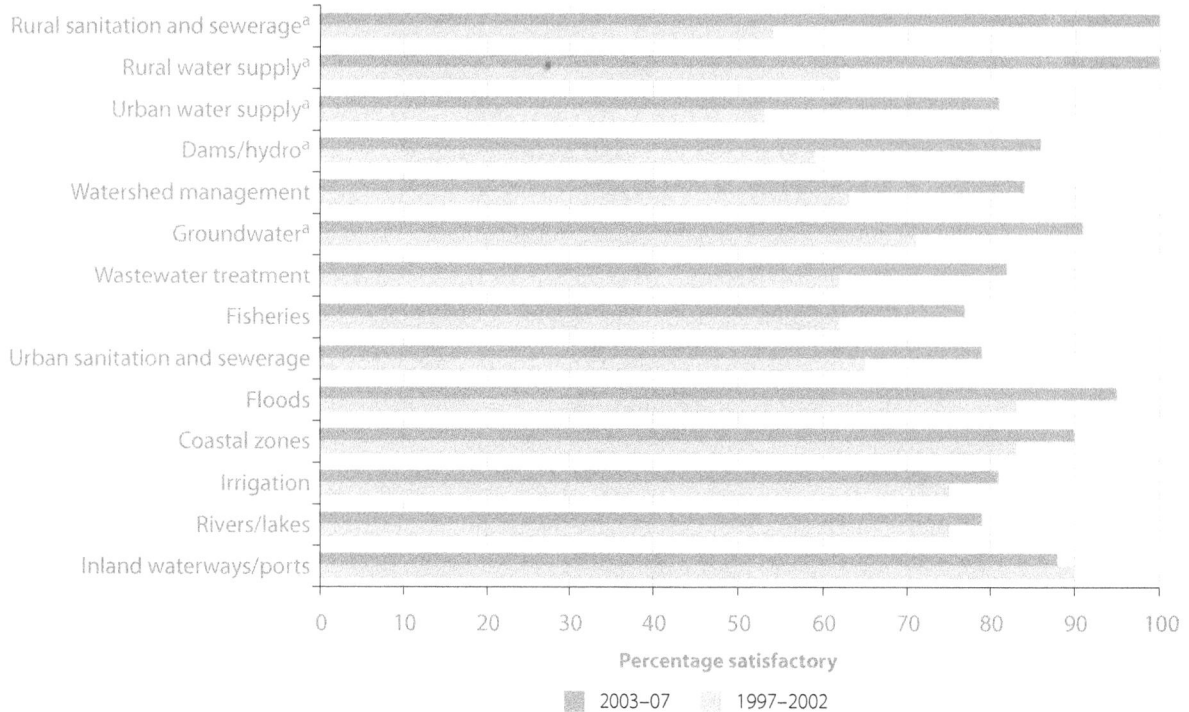

Rural sanitation and sewerage[a]
Rural water supply[a]
Urban water supply[a]
Dams/hydro[a]
Watershed management
Groundwater[a]
Wastewater treatment
Fisheries
Urban sanitation and sewerage
Floods
Coastal zones
Irrigation
Rivers/lakes
Inland waterways/ports

Percentage satisfactory

■ 2003–07   ■ 1997–2002

Source: IEG water database.

Note: Focal areas are ranked in descending order by percentage improved.

a. The difference between periods was statistically significant at the 95 percent confidence level.

FIGURE 2.6 Outcomes of Water Projects by Region

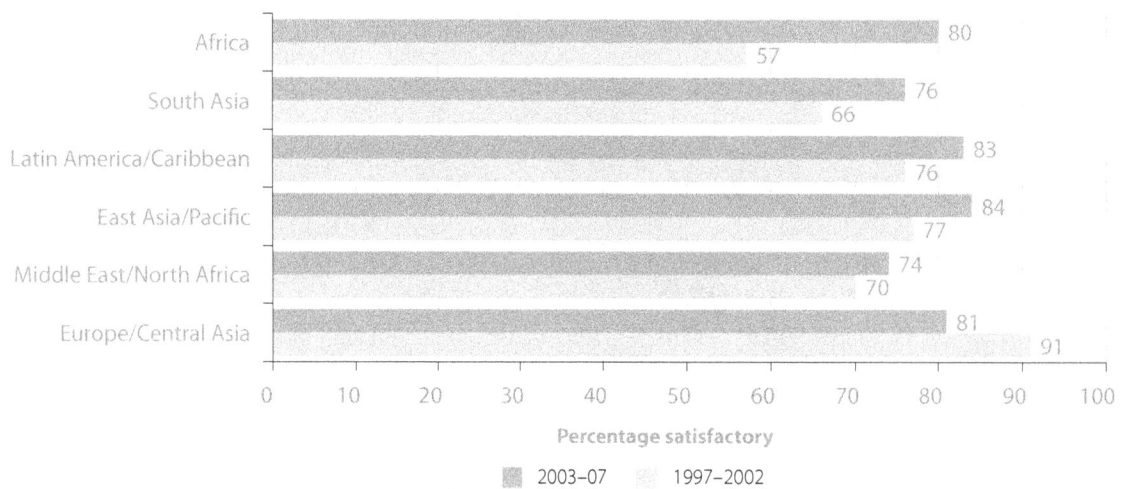

Africa — 80 / 57
South Asia — 76 / 66
Latin America/Caribbean — 83 / 76
East Asia/Pacific — 84 / 77
Middle East/North Africa — 74 / 70
Europe/Central Asia — 81 / 91

Percentage satisfactory

■ 2003–07   ■ 1997–2002

Source: IEG data.

Note: Regions are ranked in descending order by percentage improved. The difference in outcome ratings between periods is statistically significant at the 95 percent confidence level only for Africa.

water projects under the supervision of the Water Sector Board.

The WSS Sector Board managed the majority of WSS projects in the East Asia and Pacific, Europe and Central Asia, and Middle East and North Africa Regions; in the Africa and Latin America Regions, other sectors supported more projects (figure 2.7). Only in the South Asia Region were the numbers for both groups virtually the same.

A total of 253 projects dealing with WSS or with WSS components were completed and rated in this period. The ratings on all three IEG scales—project outcome, sustainability, and institutional development—were not as good for WSS Sector Board projects on average as for projects (dedicated or nondedicated) implemented by other sectors (figure 2.8). Whether the difference in ratings means that WSS activities in projects managed by other sectors produce better results than those in projects managed by the WSS Sector Board remains an open question, however, as overall outcome ratings of projects managed by other sectors may be driven by objectives and activities unrelated to WSS when WSS is only a small aspect of a project.

WSS projects implemented by the WSS Sector Board were generally rated lower than those implemented by other sector boards, but ratings for the latter may be driven by non-water-related objectives.

The lower performance of projects managed by the WSS Sector Board is due to the lower performance that characterized the WSS sector portfolio in the 1990s. Since then, the sector board has stepped up its work on underlying institutions, but financial structures, in-depth technical support, and training continue to present major challenges. Thus, the lower outcome ratings may in part reflect the more difficult challenges of water sector institutional development, which multipurpose projects that integrate WSS may not address.

The ratings of projects implemented by the WSS Sector Board have been improving substantially.

IEG's 2008 *Annual Review of Development Effectiveness* compares the performance of sector board project portfolios.[8] From 1990 to 2000, projects implemented by the WSS Sector Board had lower ratings on average than most other sectors. In recent years, however, the WSS Sector Board's ratings have shown marked improvement. In the period 1998–2002, 63 percent of water sector projects (weighted by disbursement) had satisfactory outcome ratings, compared with a Bank average of about 71 percent. For the more recent period of 2003–07, the water sector jumped to 89 percent satisfactory, bettering the Bank average by a large margin. In fact, water was the most improved sector during that period (IEG 2008a, p. 10ff).

FIGURE 2.7 WSS Projects by Region and Implementing Sector Board

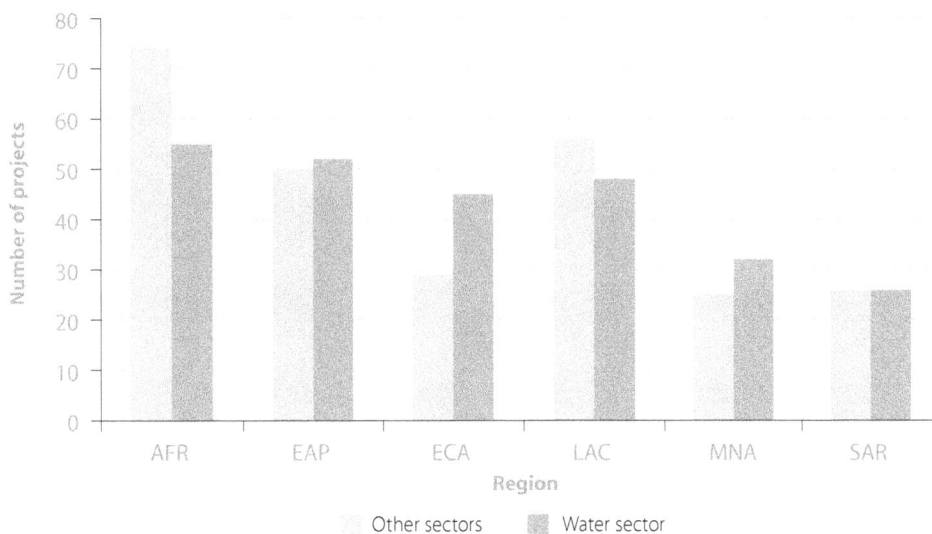

Source: IEG water database.

Note: AFR = Africa, EAP = East Asia and Pacific, ECA = Europe and Central Asia, LAC = Latin American and the Caribbean, MNA = Middle East and North Africa, SAR = South Asia.

Changes in Portfolio Focus

Figure 2.8 Outcome, Sustainability, and Institutional Development Impact Ratings for Projects Supporting WSS

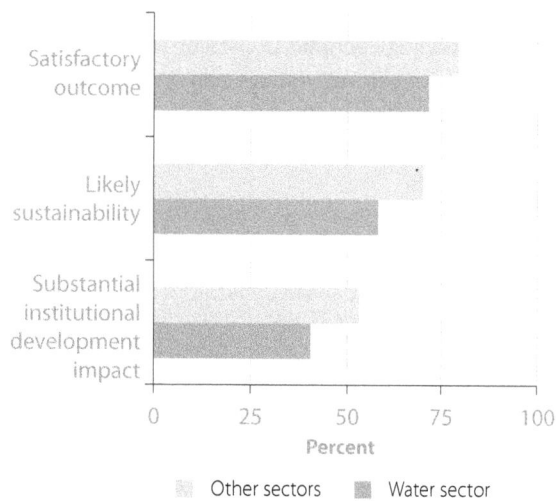

Source: IEG ratings database.

Note: Differences for sustainability and institutional development impact are statistically significant at the 95 percent confidence level; the difference for outcome was significant at 90 percent.

A closer look at the more recent trend reveals that projects (both dedicated and nondedicated) supervised by the WSS Sector Board steadily improved: satisfactory outcomes rose to 93 percent in 2005 from 63 percent in 2002; after a drop to 75 percent in 2006, the satisfactory share continued to climb, reaching 90 percent in 2007.[9] WSS Sector Board–supervised projects had a higher percentage satisfactory than did water projects implemented by all other sectors in three of the last five years (figure 2.9).

Changes in Portfolio Focus

The portfolio of 1,864 projects was analyzed in two different ways to discern the emerging priorities. First, the database was examined for the most common themes covered in the projects. These are shown in figure 2.10 (which omits 396 projects that fit best in smaller groupings not covered by the graphic).[10] Next, the percentage of ongoing projects addressing each theme as of 2007 was calculated, and these percentages were then averaged. This exercise compares the degree to which the most recent projects focus on each of the various themes. The portfolios for which this percentage exceeds the average (43 percent) are considered emerging priorities, and those that fall below the average are considered lower or diminishing priorities. For example, in 2007 attention to coastal management was 5 percentage points below the all-themes average.

Groundwater and coastal zone management account for a declining share of the portfolio, while that of dams and hydropower has been rapidly increasing.

Another way of looking at the prioritization of lending is to see how many ongoing projects address each theme in each year of the study period. Analysis of the various subsets of data reveals that attention to coastal zone management declined from 73 ongoing projects in 1998 to 51 in 2007. Other issues such as groundwater (top panel of figure 2.11) have also been declining in the portfolio. The 2003 strategy called for greater attention to major hydraulic infrastructure, and the bottom panel of figure 2.11 shows how that change in emphasis is reflected in ongoing projects year by year: since 2003 the number of dam and hydropower proj-

FIGURE 2.9 Outcome Ratings of Water Projects by Implementing Sector Board

Source: IEG ratings database.

Note: Dedicated projects only; only in 2003 is the difference statistically significant at the 90 percent level.

ects has been steadily increasing. These major themes are covered in chapters 3–6 of this report.

Prioritizing Water

Countries eligible to borrow from IDA borrow proportionately somewhat more for water (35 percent of their total borrowing from the Bank) than do IBRD countries (23 percent). However, as noted earlier, lending for water has been concentrated, with the largest share going to large IBRD borrowers: overall, 61 percent of water lending went to IBRD borrowers, and 39 percent to IDA borrowers. The big borrowers—Brazil, China, India, Indonesia, and others—typically are familiar with Bank procedures and have relatively high absorptive capacity. A relevant question, then, is whether countries that have the most precarious water situation borrow the most for water.

To answer this question, the evaluation used the Water Poverty Index (WPI) constructed by the United Kingdom's Natural Environment Research Council, Centre for Ecology and Hydrology, using data gathered by the World Resources Institute. The WPI measures, for a given country, the impact on human populations of water scarcity and the current level of water provision.[11] Each WPI is an average score for the country as a whole and conceals intracountry variation, which can be substantial. The WPI ranges between 0 and 100, where a low score indicates water poverty and a high score indicates good water provision. The WPI for each Bank borrower was then compared with the amount per capita of Bank lending for water to that country (figure 2.12).

The comparison shows no clear relationship between Bank water lending and water stress: the correlation coefficient is about 0.2. Similarly, when each country's allocation for water within its overall borrowing is compared with its water poverty ranking, no relationship is found.[12] This is not to say that the Bank should stop providing support to water-rich countries. And, of course, a single large hydropower project would quickly change the picture for a given country. (Country-by-country data on water borrowing can be found in appendix table J.11.)

Bank lending for water is not correlated with the Water Poverty Index.

The Bank's water strategy covers all countries without prioritization. The Bank's 2003 strategy paper says:

- In all countries there is a major need for more effective management of water resources….

- In all countries there is a need for greater attention to water allocation, demand management, water rights and the use of pricing and other economic instruments….

- In all countries there is a need for improving the benefits from existing infrastructure and for developing institutional and financial arrangements for sustainable rehabilitation and maintenance [World Bank 2003b, p. 41].

One way the Bank might help borrower countries (especially IDA borrowers) would be to make a greater effort at

---

FIGURE 2.10 Shares of Ongoing Projects Addressing Selected Themes

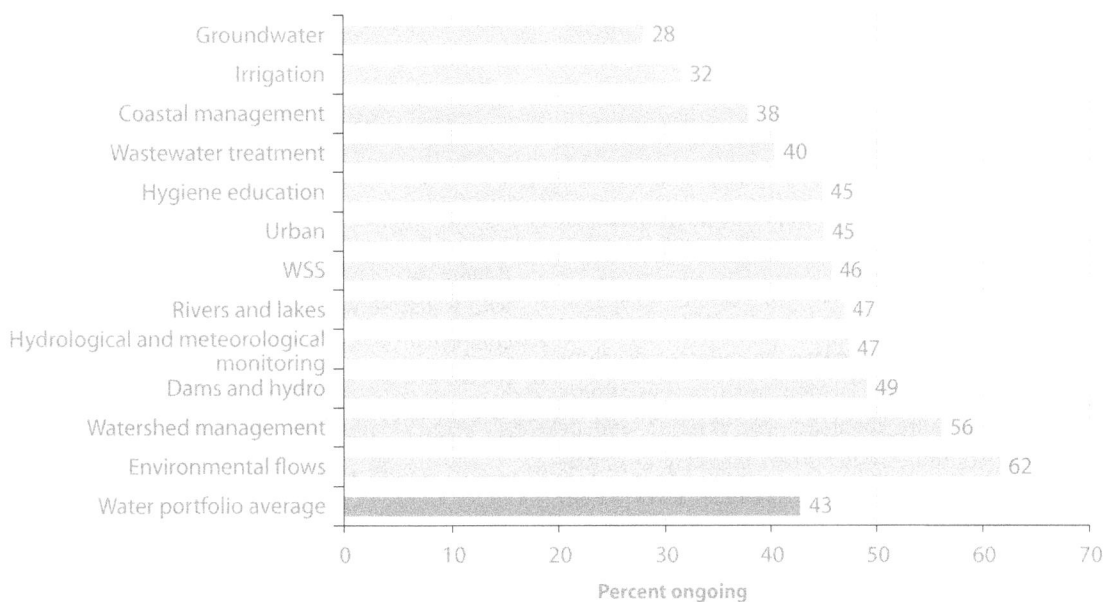

Theme	Percent ongoing
Groundwater	28
Irrigation	32
Coastal management	38
Wastewater treatment	40
Hygiene education	45
Urban	45
WSS	46
Rivers and lakes	47
Hydrological and meteorological monitoring	47
Dams and hydro	49
Watershed management	56
Environmental flows	62
Water portfolio average	43

Source: IEG water database.

Groundwater Projects

Dam and Hydropower Projects

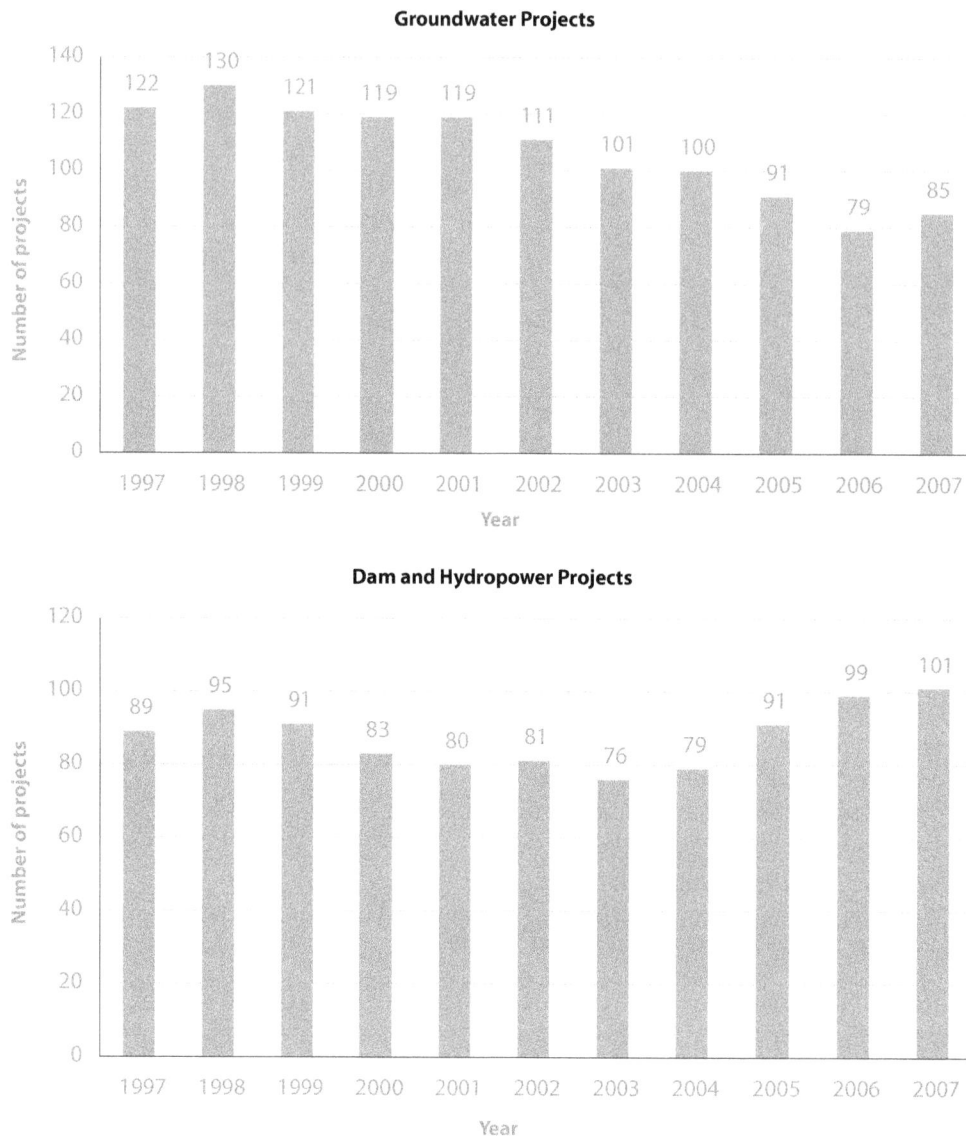

Source: IEG water database.

the design stage to find the country-specific institutional arrangements, policies, and investments that will make a difference. All of these need to be compatible. The Bank could help by producing more Country Water Resource Assistance Strategies that review the different developmental, economic, and social returns to various allocation scenarios in more countries—for example, and by helping countries under water stress integrate this challenge even more fully into their national development strategy.

The Bank's water strategy does not single out water-stressed countries.

Summary

A large part of what the Bank finances has something to do with water. Between 1997 and 2007 the Bank approved or completed 1,864 projects and grants with at least one water-related activity. Water-related projects represent 31 percent of all Bank projects approved since 1997, and 28 percent of Bank commitments. Water lending grew in both commitments and number of projects during the period.

The integration of water practice across Bank sectors—an important goal of the Bank's water strategy—appears to be well under way. The majority of water-focused projects are overseen by sector boards other than that for water. The Ru-

Bank water commitments per capita in U.S. dollars

Water Poverty Index

Water-rich countries

Water-poor countries

Water Poverty Index

Bank water commitments per capita in U.S. dollars

Sources: World Resources Institute and IEG data.

ral Sector Board is responsible for the largest share of water-related lending and has supervised the most projects.

Performance of Bank-funded water projects is on par with the rest of the Bank's portfolio and has been on an improving trend. In general, the best performers have

Photo courtesy of Curt Carnemark/World Bank.

been waterways and ports and flood reconstruction. Urban water supply, sewerage, and sanitation performed worse on average than other water-related projects. During the first half of the evaluation period, projects managed by the WSS Sector Board had lower ratings, on average, than did projects in most other sectors. In recent years, however, ratings of WSS Sector Board–managed projects have shown marked improvement. The sector board has stepped up its work on underlying institutions and financial structures, in-depth technical support, and studies and training, although, as will be seen in subsequent chapters, much remains to be done on institutional development.

There is no apparent correlation between a country's water stress and Bank lending for water to that country. To ensure that water issues are adequately addressed in countries facing severe water stress, the Bank should look for entry points to help countries make water use more sustainable, even if the Bank may not necessarily be able to finance all the work that is needed to resolve the most pressing water issues.

Chapter 3

EVALUATION HIGHLIGHTS

- Combining livelihood improvements with environmental restoration in watershed management projects generally produced good results.

- Despite the seriousness of groundwater depletion, the extent of the groundwater problem is poorly understood, and few projects protect or restore aquifers.

- The Bank has usually succeeded in establishing river basin management organizations, but their sustainability has been an issue.

- Monitoring has been supply driven, and projects often did not identify who would use the data collected.

- Limited success with full cost recovery has caused the Bank to moderate its approach.

- Water efficiency initiatives are undertaken most often in water supply and irrigation and drainage projects, where they commonly promote the use of improved technology, reduction of losses, and institutions for better management.

- As groundwater has become increasingly scarce, ongoing Bank projects have shifted away from investments in extraction.

Photo courtesy of Curt Carnemark/World Bank

Managing Water Resources

The Bank helps developing countries manage their water resources through loans, credits, and grants that build infrastructure for or support the management of watersheds, river basins, and groundwater, or that support hydrometeorological monitoring.

This chapter explores the variety of ways in which the Bank has promoted water resource use that takes into account the currently available supply and the likely future demand. As this is largely a story about sustainability, the chapter also discusses the ways in which Bank-financed projects have tried to manage the demand for water.

Watershed Management

By using the drainage basin or catchment area as the unit of intervention, the watershed management approach usually aims to help improve upland natural resources management in order to protect downstream resources and infrastructure. A watershed can be thought of as a geographical area that drains into a single river system, and a river basin as the river itself and all its tributaries. Hence, watershed management is more concerned with how water affects the land on its way to a body of water, whereas basin management has more to do with how the water is used once it reaches that body of water.

The Bank defines watershed management as "the integrated use of land, vegetation and water in a geographically discrete drainage area for the benefit of its residents, with the objective of protecting or conserving the hydrologic services that the watershed provides and of reducing or avoiding negative downstream or groundwater impacts" (World Bank 2008e, p. xi).

Therefore, watershed management projects are designed to overcome obstacles to environmentally sensitive natural resources management by incorporating the interests of stakeholders (and often the provision of alternative livelihoods—a practice termed the "livelihood-focused approach") in technological solutions to the typical problems of watersheds.[1] The problem these projects often face is that in most watersheds, people making a living from degraded lands generally contribute to the degradation with their daily activities, and the people causing the damage have to become part of any solution.

IEG identified 218 projects in its water database with at least one watershed management activity; these projects had total commitments of $13.1 billion over the 10-year evaluation period. Most projects concentrated on land manage-

ment improvements that affect surface water. The size of the treated watershed varied considerably across projects (appendix table J.12). Eighty-five projects (39 percent) used the approach promoted in the Bank's 2003 Water Resources Sector Strategy, combining livelihood improvements with environmental restoration. Projects that took a livelihood-focused approach to watershed management performed far better (90 percent had satisfactory outcomes) than those that did not (69 percent; figure 3.1).

Across all completed watershed management projects, those that took a livelihood-focused approach performed better than average.

To better understand these results, IEG analyzed the 31 completed projects that focused on livelihoods. In the 25 projects with data on beneficiaries, a total of about 9 million farmers—most living in extreme poverty—benefited from the project (appendix table J.12). Families and communities received wells, water storage facilities, rural roads, improved houses, rural electrification, and social infrastructure. In return for the benefits associated with the construction of infrastructure (such as irrigation systems), local communities had to commit to environmental restoration activities (box 3.1). Seventy-four percent of livelihood-focused projects paid special attention to women, because their activities were more closely involved with natural resources management.[2] Forty-eight percent of projects also focused on minorities and tribal areas, such as Bedouin families in the Arab Republic of Egypt.

Three-quarters of livelihood-focused projects paid special attention to women.

To increase the incomes of the local population, productive capacity was improved through market gardens, livestock improvement, veterinary outreach, small irrigation schemes, grain storage facilities, and mills. These income-enhancing activities proved successful in all but two cases (Brazil and Uruguay). In those cases, circumstances exter-

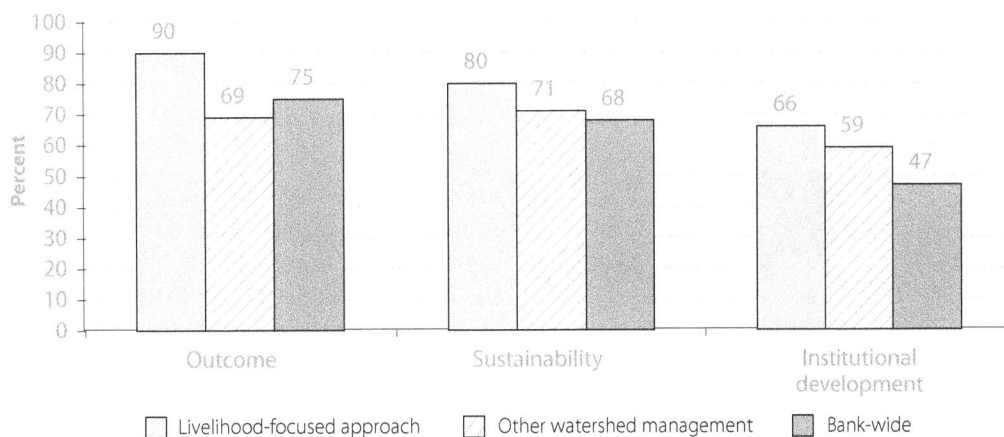

Source: IEG water database, based on the full universes of relevant projects.

nal to the project (drought and macroeconomic conditions) best explain the disappointing results.

Watershed management projects that included livelihood interventions were often claimed to have a low benefit-cost ratio because environmental benefits were not quantified.

Although watershed management projects combining livelihood interventions with environmental restoration enjoyed high success rates (appendix figure J.9), project completion reporting often described the benefit-cost ratios as low, both because environmental benefits were not taken into account and because there was a time lag between action and downstream benefit in some watersheds. The Morocco case study carried out for this evaluation provides an example (appendix box J.1). Effects on downstream communities, such as reduced siltation of reservoirs, reduced flooding, and improved water quality and availability were rarely measured, nor were the social and organizational benefits from averting potential conflicts between upstream and downstream communities taken into account. Hydrological monitoring

BOX 3.1

IMPROVED WATER MANAGEMENT PRACTICES

In the Integrated Watershed Development (Plains) Project (P009860, completed in 1999), Indian farmers on private lands benefited directly from water-controlling and -harvesting structures built in natural drainage lines in lower parts of the watersheds. Earthen runoff structures, usually constructed at the downstream end of gullies, were very popular among farmers in all project areas. These structures not only provided additional drinking water to animals during dry spells but have assisted in the recharge of groundwater, increased the soil moisture of downstream fields, provided water for supplemental irrigation on private land, and controlled runoff water and soil erosion. Concrete and rock diversion structures were also constructed on natural drainage lines and seasonal streams. This resulted in an increased agricultural water supply for adjacent fields, leading to improved harvests of rice and other field crops, vegetables, and fruits.

Source: ICR for the India: Integrated Watershed Development (Plains) Project.

Photo courtesy of Dominic Sansoni/World Bank

(with or without remote sensing) and watershed modeling might have helped improve impact assessment and thus have better captured the true benefit-cost ratio of such interventions. One GEF project visited by IEG demonstrated how this approach could work (box 3.2).

Groundwater Management

Groundwater is increasingly threatened by overexploitation, inadequate environmental flows, and contamination. Throughout the world, water from rivers and reservoirs has been committed to household consumption and to economic uses without adequately considering the need to replenish groundwater or sustain aquifers or to protect water quality by safely returning used water to the environment (Clarke and King 2004, p. 52). Groundwater depletion is most severe in the Middle East, North Africa, and South Asia (box 3.3). In some coastal areas so much freshwater has been withdrawn from aquifers that salt water has started to intrude, threatening the usability of what remains. This problem was encountered in the Morocco, Tanzania, and Vietnam case studies and is discussed further in the section on coastal zones.

> As groundwater has become increasingly scarce, ongoing Bank projects have shifted away from investments in extraction.

Over the evaluation period as a whole, extractive activities, such as construction of groundwater supply schemes or well construction for irrigation, dominated Bank-supported

GROUNDWATER IS DEPLETING RAPIDLY IN THE REPUBLIC OF YEMEN

The Republic of Yemen has no significant perennial sources of surface water. Instead it relies almost exclusively on exploitation of groundwater. Water is taken from the shallow aquifers, which are rechargeable, and increasingly from deeper aquifers, which are generally considered not rechargeable, although some recharging has been accomplished with significant difficulty and expense. In large parts of the country, water from the shallow aquifers is extracted at well over the recharge rate from the country's limited rainfall. Thus, as water from these aquifers is exhausted, pumping relies on the deep (fossil) aquifers that are also depleting. Because these aquifers cannot be readily recharged, pumping is essentially a mining operation. Groundwater tables are declining inexorably in many locations, sometimes at dramatic rates. In the Sana'a Basin, the groundwater table is falling by as much as 6 meters per year in the more populated areas, and rural and urban tube wells are constantly being deepened.

The rate of groundwater depletion has accelerated over the past three decades for several reasons. One has been the explosion in agricultural use resulting from the introduction and rapid spread of mechanized tube wells. Farmers, using shallow dug wells, had traditionally extracted water at about the rate of natural recharge from rainfall, but this changed when the tube well technology and pump-sets were introduced in the early 1970s. This enabled much higher levels of groundwater abstraction as well as pumping from the deep aquifers. Agricultural use of water increased by about 5 percent per year in the 1990s. In 1990 agricultural consumption (for irrigation) was 130 percent of the country's renewable water resources; it has since increased to 150 percent.

Source: IEG Republic of Yemen case study research.

groundwater projects (appendix figures J.9 and J.10). However, as groundwater has become increasingly scarce, ongoing Bank projects have shifted away from investments in extraction (figure 3.2). For example, the number of projects financing new drilling for irrigation has declined. The Morocco case study research found, however, that in the absence of a regulatory system and systematic enforcement, many small-scale private providers step in to provide—often illegally—access to groundwater for those with the capacity to pay.

The number of projects dealing with groundwater conservation also declined somewhat after 2001. This set of projects briefly witnessed a small positive shift toward increased sustainability and conservation before dropping below its original level (figure 3.2). Relevant activities include monitoring of groundwater quality, landfill site improvements to reduce leachate, and the reduction of infiltration of contaminated surface water into groundwater (appendix table J.13). As box 3.4 reports, aquifer recharge, a proven technology,[3]

FIGURE 3.2 Ongoing Groundwater Projects Focusing on Extraction and Conservation

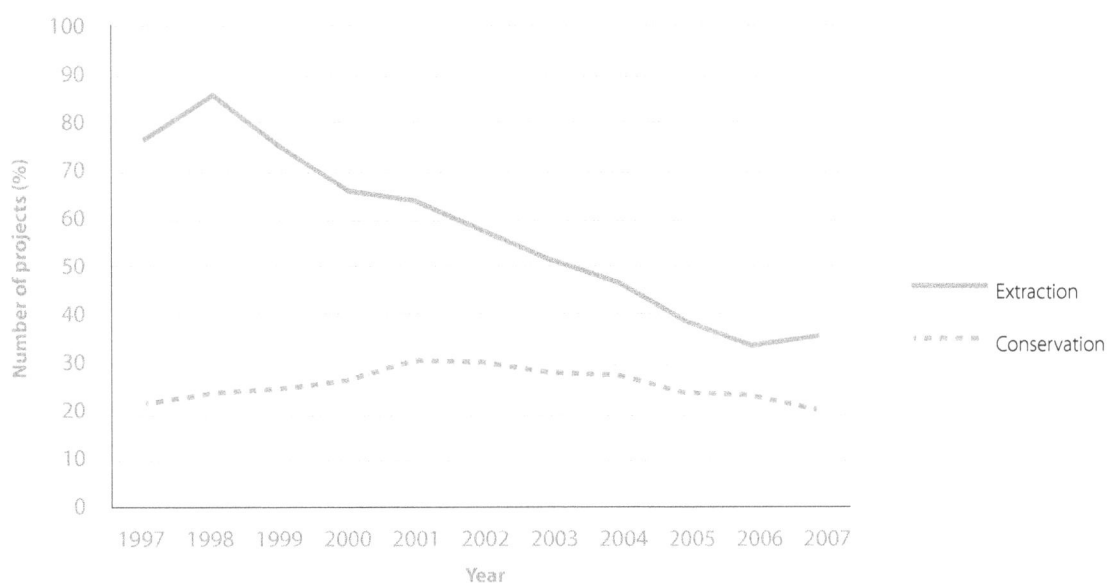

Source: World Bank data.

ARTIFICIAL RECHARGE OF THE SOUSS AQUIFER, MOROCCO

The water table is falling dramatically in the Souss Aquifer near Agadir in southern Morocco. In recent decades farmers have had to deepen wells every two years, because water tables were falling by 1.5 to 3 meters per year. Under the Water Resources Management Project (P005521), five weirs were reconstructed and one new one built to prevent dam overflow from being lost to the sea. Contrary to expectations, water does not infiltrate the entire aquifer, although it is detectable quite far from the insertion point—in this case it could still be measured more than 80 kilometers from the dam. The weir does not help communities that have to tap into its groundwater from farther away, but the objective of mobilizing additional groundwater by artificially recharging the Souss aquifer was met in the limited area where it was attempted.

Source: IEG Morocco case study research.

helps restore partially depleted aquifers, but there are only a handful of Bank-financed efforts of this type worldwide. To what extent this technology is more consistently supported by other donors is not clear.

Analysis of the successes and failures of various groundwater-related activities in completed projects found that five activities had success rates at the 90 percent level or above (figure 3.3). Construction of groundwater supply schemes not only was attempted most frequently, but achieved the intended goal most often. In general, activities whose objective was increasing water supply were the most successful. This result is most likely a function of the simplicity of the goal—get more water—as well as of strong stakeholder demand and concomitant close oversight.

The least successful activity in the groundwater portfolio was the development of management frameworks or plans: only 20 percent of projects attempting this activity were able to complete the task. Several activities in the bottom 10 are related to reducing pressure on groundwater and to conservation: expanding or establishing surface water sources in place of groundwater; groundwater recharge schemes; and activities for the development of alternative sources, such as rainwater harvesting.

Several findings stand out when comparing the most and least successful groundwater-related activities. First, the more successful activities tend to be not particularly challenging by nature. They are focused on the construction of infrastructure and the extraction of groundwater. In contrast, the least successful projects address environmental and resource protection issues critically important to the safe, longer-term use of groundwater resources. Although the more successful activities include those related to groundwater monitoring, two of the least successful activities are related to improvement in the quality of water through pollution abatement. Particularly troublesome is

FIGURE 3.3 The 10 Most Successful Activities Dealing with Groundwater

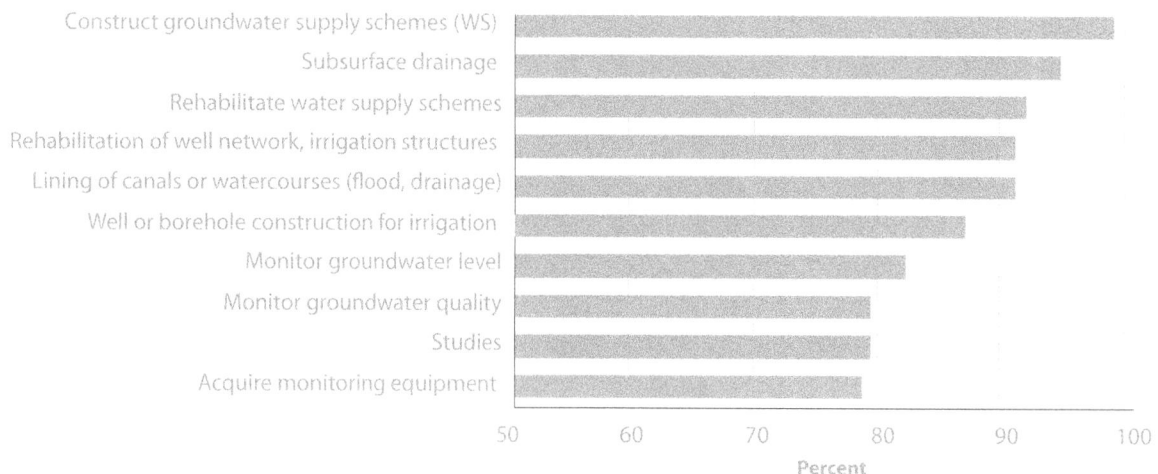

Source: IEG water database.

Note: Each bar reports the success rate in attaining groundwater-related goals within the indicated activity.

FIGURE 3.4 The 10 Least Successful Activities Dealing with Groundwater

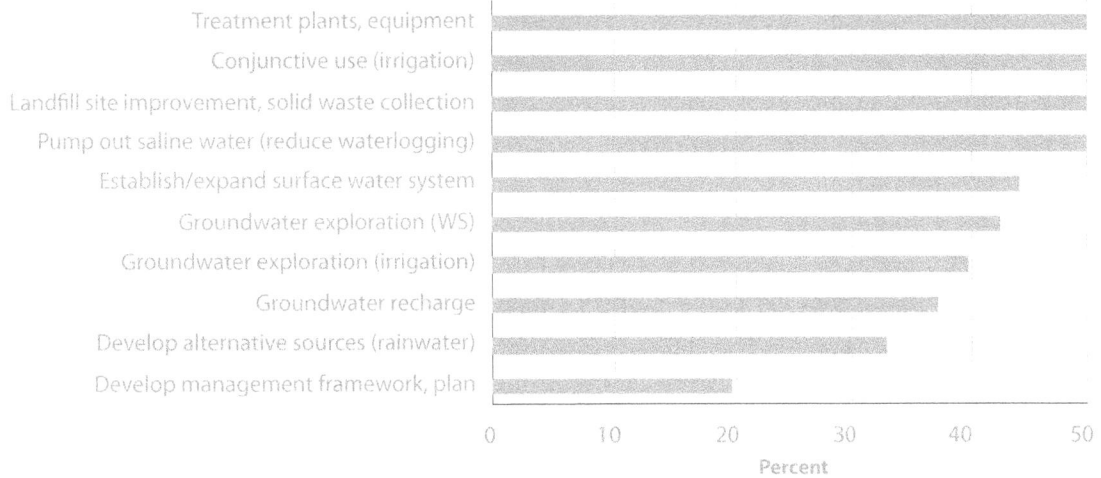

Bar chart with vertical axis labels (top to bottom) and horizontal axis "Percent" from 0 to 50:

- Treatment plants, equipment
- Conjunctive use (irrigation)
- Landfill site improvement, solid waste collection
- Pump out saline water (reduce waterlogging)
- Establish/expand surface water system
- Groundwater exploration (WS)
- Groundwater exploration (irrigation)
- Groundwater recharge
- Develop alternative sources (rainwater)
- Develop management framework, plan

Source: IEG water database.

Note: Each bar reports the success rate in attaining groundwater-related goals within the indicated activity.

that the least successful activity of all—the development of groundwater management frameworks and plans—is critical to sustainability.

Bank projects need to be more ambitious in addressing issues critical to the long-term use of groundwater.

The extent of groundwater depletion is poorly understood because groundwater data are rarely collected and even less often publicly available. Research into the condition of groundwater supplies has had a low priority, and many countries do not have enough information at hand to make long-term plans. Within the set of projects examined by this evaluation, it was quite common for project appraisals to state an intention to monitor groundwater quality and level, but these intentions often had not been carried out by the time the loans closed. With many areas experiencing changing rainfall patterns and climate, the hydrological record has been a poor predictor of the future. This means

that the collection and analysis of groundwater data are more important now than ever and need to be taken on board more commonly than they have been. To this end, the Bank has recently established a Groundwater Management Advisory Team, cofinanced by DFID and the Bank-Netherlands Water Partnership Program.

The extent of the groundwater problem is poorly understood because data are rarely collected or shared.

River Basin Management

Increased stresses on a river basin's natural resource base are felt well beyond the immediate area of a particular intervention. Conversely, even small river basin improvements can have big payoffs (Dyson and others 2003, p. 17).

Even small river basin improvements can have big payoffs.

The 1992 Dublin Principles suggest that independent management of water resources by different water-using sectors in a river basin is suboptimal, because there is no one to take the water resource needs of all users into account and to balance them in a sustainable manner. The Bank's strategy supports this stance. Sound river basin management requires an understanding of how much water can be consumed while still meeting environmental flow requirements (without overexploitation of groundwater). How best to apply the Dublin Principles depends, of course, on the location.

One way to support an integrated approach to river basin management is through river basin organizations (RBOs). These are part of government, but if they are to have authority to make allocative decisions and reduce water stress, they need significant political support. Basin management is anchored in the subsidiarity principle—tasks should be undertaken by the most localized entity competent to do so—but its actual application often encounters resistance from groups slated to lose power or privilege (World Bank 2005a, 2006b).

An RBO's form and role need to reflect the historical context and interaction patterns among critical stakeholders; this varies by basin, as can be seen in appendix table J.14. Generally, however, the key characteristics of an effective RBO include the following (World Bank 2006b):

- Basin-wide planning
- Balancing of all user needs for water resources
- Protection from water-related hazards
- Wide public and stakeholder participation, with attention to gender and minority group issues.

BOX 3.5

RIVER BASIN MANAGEMENT IN TANZANIA

Water security remains elusive in Tanzania, despite an endowment of freshwater resources that is relatively large by global standards. Water insecurity is characterized by the vulnerability of people and the economy to frequent cycles of drought and flood, increasing conflicts over water, and increasing degradation of water resources. Recently, scarcity and conflict among users have been exacerbated by a few years of below-average rainfall, coupled with the expansion of irrigation and an increase in demand for hydropower and water for urban uses.

Along with local conflicts within an irrigation network have come conflicts over the upstream versus downstream uses of rivers. After consultations with the World Bank and a rapid water resources assessment undertaken by Tanzanian water experts, the government decided to improve water resources management through the establishment of basin organizations and the adoption of an IWRM approach.

Although promulgated by law in 1981, the new organizations, called basin water offices (BWOs), became operational only in the 1990s, the first one being the Pangani Basin Water Office in 1991, followed by the Rufiji Basin Water Office in 1993. The River Basin Management and Smallholder Irrigation Improvement Project (P038570) provided ongoing support to the fledgling organizations. The project provided financing for the necessary premises, computer devices, gauging stations, and vehicles. As of 2008 these two BWOs were substantially operational, while the other seven are being supported under an ongoing Water Sector Support Project (P087154).

One of IWRM's objectives for river basin management is to ensure that each entity using water (irrigators, herders, industries, hydropower, fisheries, national parks, mining, urban and rural water supply and sanitation, and environment) is represented in decision making. During a 2008 IEG workshop with representatives of the Pangani BWO, participants acknowledged that the basin office had so far neglected the needs of herders and the environment. A promising recent development is that the Pangani BWO has completed a basin-wide environmental flow assessment that takes a more inclusive look at all stakeholder needs and maps out a road to a better environmental outcome.

Source: IEG Tanzania case study research.

RBOs oversee such concerns as water allocation, water regulation, resource management and planning, education of basin communities, development of natural resources management strategies, and programs to remediate degraded lands and waterways. They may also contribute to consensus building and conflict management, as was found in the evaluation's case study research in Tanzania (box 3.5).

The review of the portfolio found that 30 projects worked with basin management institutions. Some of these established new RBOs, while others tried to strengthen existing ones. (For details see appendix table J.14.) Eleven basin management projects closed during the period studied and were analyzed for their results. Figure 3.5 shows, for several identified RBO activities, the share of projects that achieved positive results and the share that did not.

Bank support to help establish new RBOs was generally successful, but their sustainability remains in question.

The results suggest that the Bank has generally succeeded in helping to establish new RBOs, but the sustainability of these organizations often remains in question. It is quite common in development for donor priorities and funding to induce stakeholders to create new organizations. If the objective calls for the creation of such an organization, the fact of its creation can be labeled a success. But institutional sustainability is an important parallel issue. In the absence of postproject government support and the retention of carefully trained staff, such organizations often wither. The review found that in seven out of eight cases institutional

creation proceeded as planned.[4] Nevertheless, nine projects stated an intention to strengthen basin organizations, but only two reported positive results. In five other cases the basin management agencies were reported to be too weak to function as intended, and by project closing the RBOs still had limited access to financial sources and technical assistance. Completion reports for two projects reported that the basin agencies established were on the verge of collapse for lack of sufficiently skilled and motivated staff.

Hydrological and Meteorological Monitoring

Bank-supported hydrological and meteorological monitoring projects show a shift in emphasis toward more local and regional networks.

Hydrological and meteorological monitoring systems provide weather and water-level data on rivers, lakes, reservoirs, and groundwater. This information can help to prevent natural catastrophes, determine the volume of water flows, and assist in water resources management decisions. Technological advances in sensor technology and the integration of electronics and data communication have made the automation of meteorological and hydrological networks increasingly affordable and attractive to policy makers and planners (Kokko and Vaisala 2005).

The set of Bank interventions concerned with hydrological and meteorological monitoring consists of 55 projects. Over the period studied, the Bank helped countries to in-

FIGURE 3.5 Selected Results of Work with River Basin Organizations in Bank-Financed Projects

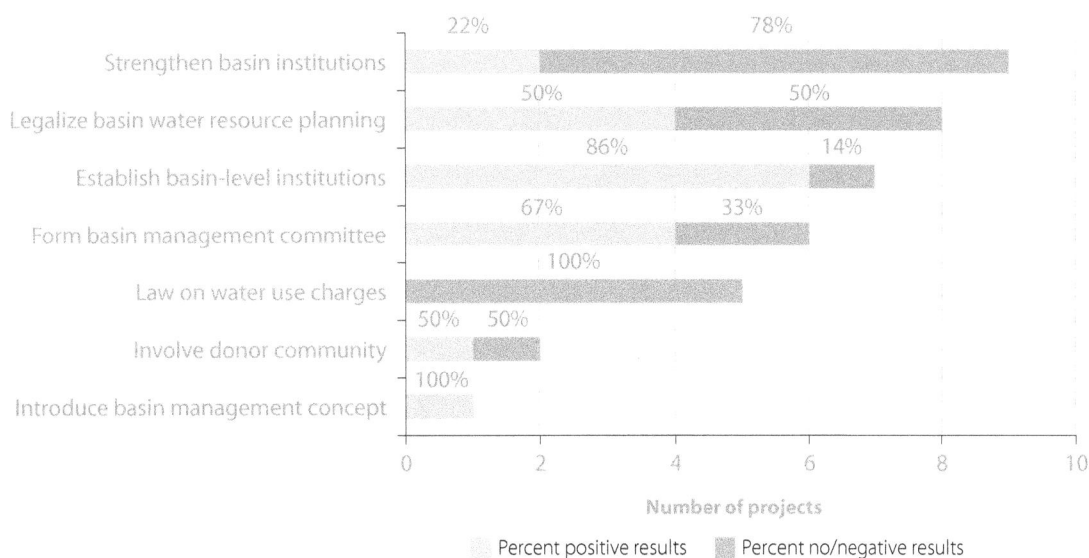

Source: IEG water database.

EARLY WARNING SYSTEM USED FOR MONITORING LAKE SAREZ IN TAJIKISTAN

In Tajikistan nearly two-thirds of the population rely on agriculture, but many live in areas particularly vulnerable to natural disasters. In the mountainous province of Gorno Badakhshan, although many small, remote villages are prone to earthquakes and seasonal flooding, the greatest danger is posed by Lake Sarez.

The lake was formed in 1911 after an earthquake when a 2.2-billion-cubic-meter landslide blocked the Murghab River Valley. The landslide buried the village and the inhabitants of Usoi, giving the landslide dam its name. Reports on the Usoi dam describe it as being of questionable stability. These reports suggest that if a strong earthquake should occur in the vicinity of the lake, another landslide could fall into the lake, generating an enormous wave that could overtop the natural dam, possibly washing it away. The resulting flood could affect up to 5 million people living in surrounding areas. The Bank-financed Lake Sarez Risk Mitigation Project (fiscal 2000) was designed to prepare and protect vulnerable settlements in the event of such a flood; the project also includes activities related to the mudslides, rock falls, avalanches, and seasonal flooding that are common in the project area. The project installed a monitoring system and an early warning system essential to prevent a catastrophe. The installed monitoring system at Lake Sarez allows for the collection of hydrological, meteorological, and seismological data in the lake area and enables the monitoring of the most rockslide-prone bank and of the Usoi dam itself. The major advantage of the system is its automaticity, which reduces response time and allows appropriate decisions and timely alerts in case of emergencies.

Sources: World Bank project documents (ICR for project P067610).

vest in four types of monitoring systems: early warning systems (21 projects; box 3.6),[5] combined hydrological-meteorological monitoring systems (21 projects), free-standing hydrological monitoring systems (18 projects), and free-standing meteorological monitoring systems (19 projects). Some projects involved more than one type of system.

An activities analysis (appendix table J.16) found that developing monitoring systems at the national level has been the most common approach to area coverage: the 23 national systems projects (15 closed, 8 active) accounted for 42 percent of the total of 55 projects (figure 3.6). There has been a shift in emphasis, however, with the Bank now financing more local and regional networks. Projects financing transboundary monitoring systems have also increased in number.

A sizeable percentage of monitoring projects did not tailor systems to meet beneficiary needs.

The achievements of monitoring activities were analyzed for the 28 closed projects that had supported hydrological

FIGURE 3.6 Hydrometeorological Monitoring Systems by Level of Supervision

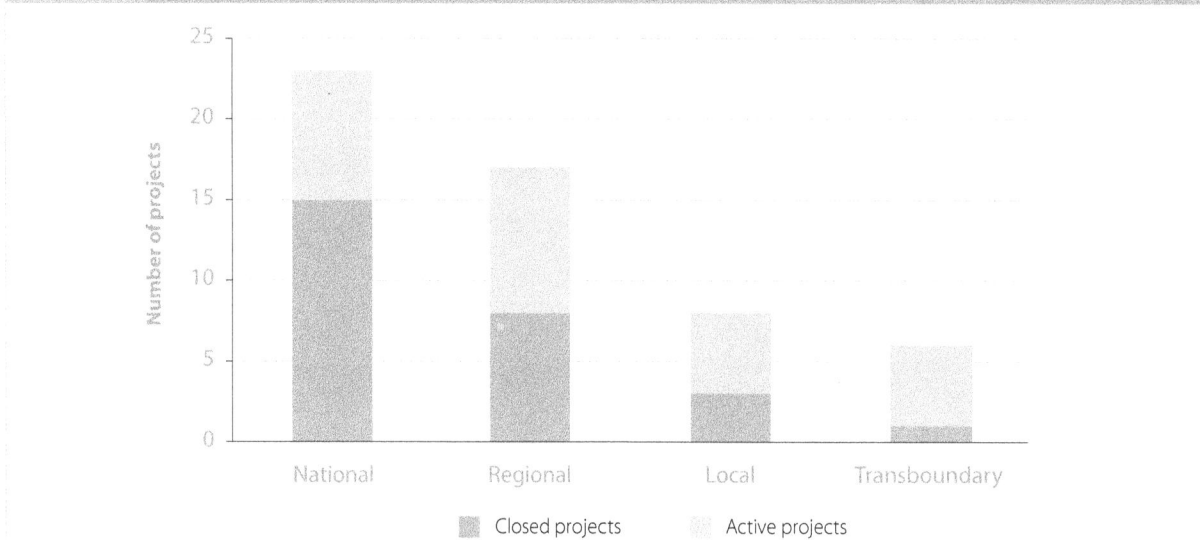

Source: IEG water database.

monitoring (appendix table J.18). In general, this type of monitoring has been supply driven. The evaluation found that 43 percent of projects clearly did not manage to tailor systems to meet beneficiary needs, and most of the rest did not address the issue of who is to use the data. Instances where the project files indicate that systems benefited stakeholders and improved decision making in water resources management and natural disaster mitigation were not common. Over half of the projects (54 percent) that used monitoring data for disaster prevention and mitigation succeeded in getting the information into the hands of people whose job involved mitigating natural disasters and reducing damage. But in 43 percent of projects there were no clear links between collecting the data and using it to take corrective action, make changes in practice, or inform policy. In the absence of practical utility, a sustainable process was rarely established (box 3.7). Weaknesses in monitoring systems were usually due to deficiencies in project design, especially with regard to stakeholder participation, maintenance, and the appropriate choice of monitoring equipment and facilities. In a fast-changing world, not having the information at hand to adapt to a rapidly changing climate will make water management situations increasingly problematic.

Training for O&M has been hindered by its short-term nature, by hiring freezes and staff size restrictions, and by a worldwide shortage of hydrological specialists.

The provision of equipment and operation and maintenance (O&M) training were the two most frequently implemented activities in hydrological and meteorological monitoring projects (see appendix table J.17), but the training given often focused on teaching how to install and operate the equipment, and not so much on how to interpret and present the data. Failure to fill key positions and staff size restrictions were reported in 18 percent of projects; these, too, are obstacles for the O&M of monitoring systems and programs. Finally, because of a worldwide shortage of specialists, demand for monitoring skills in the private sector led to difficulties in acquiring and retaining staff during or after project closure.

Managing Efficiency of Use and Demand for Water

Because the supply of water is finite and pressures on that supply are increasing rapidly, managing the demand for water will be critical to future development. Demand for water can be affected by three broad sets of interventions: pricing, quotas, and measures to improve water use efficiency. Demand-side management (DSM) is defined as any attempt to encourage water users to control their use. Increasing water prices (or tariffs) in line with marginal costs can help to reduce water demand and encourage use of the resource commensurate with its scarcity. Pricing interventions may be effective in reducing water use for domestic water supply and energy generation but are less so for reducing agricultural water use, which responds less to changes in price. For this reason the Bank has started to encourage borrowers to introduce quotas to manage agricultural water consumption.

Measures to enhance the efficiency of water use range from investments in technology (such as water-efficient toilets) to reductions in unaccounted-for water (UfW). Demand management and water pricing interventions have to go hand in hand with infrastructure developments in responding to water demand. In addition, consciousness raising for service providers (agencies and water user associations, or WUAs) is needed to increase their motivation to provide

BOX 3.7

SUSTAINABLE MONITORING IS DIFFICULT TO ESTABLISH

As of 2008, Tanzania did not have a functional network for monitoring water resources and flows. The existing system did not systematically collect, integrate, and analyze hydrological data at the central level. Political support for data collection was low—priority was given instead to providing a supply of potable water—and during the IEG visit each of the two agencies responsible for the system attributed its dire condition to the other. The IEG team found that the electronic data processing equipment that had been procured required skills that the operators did not possess, and spare parts for fragile equipment damaged accidentally were not locally available.

The monitoring system set up for Morocco's Oum-er-R'bia Basin provides an interesting contrast with Tanzania's approach. Morocco decided not to use a real-time data gathering system, because set-up and operating costs were too high. But whereas Tanzania's higher-tech system ultimately did not work, Morocco's approach resulted in a robust system and an impressive database (but no real-time data). The contrast highlights the importance of tailoring each monitoring system to local conditions and purposes.

Sources: Tanzania and Morocco case study research.

water users with better service, even when this involves the imposition and enforcement of consumption quotas.

Although the promotion of DSM is not a specific concern of either the 1993 or the 2003 water resources strategy, about a quarter of all water projects in the evaluation (539 of 1,864) deal with some aspect of DSM:[6] 321 projects undertook discrete activities to increase the efficiency of water use (figure 3.7); 143 aimed to increase water tariffs to encourage users to reduce water use; and 141 projects planned to reduce UfW. After a peak in the late 1990s, the number of annual approvals of DSM projects declined (appendix figure J.12).

Improving Water Efficiency

Water efficiency initiatives are found most often in water supply and irrigation and drainage projects.

The evaluation analyzed the water portfolio subsector by subsector, identifying those that promote water efficiency (see table 3.1). The largest group was implemented under the rubric of water supply (27 percent). The next most efficiency-concerned subsector was irrigation and drainage (18 percent).

Water Use Efficiency in Agriculture

Irrigation and drainage projects commonly promote the use of improved technology, reduction of water losses, or the creation of institutions for better management.

Efforts to improve efficiency of water use among farmers are common, which is understandable considering that agriculture is the biggest user of water. These efforts are usually focused on increasing the value of agriculture production per unit of water consumed through increasing yields and reducing nonbeneficial water use (such as losses from seepage). Alternatively, projects may promote alternative crops to irrigators to reduce demand on the irrigation system. Ninety-seven irrigation and drainage projects made some attempt to increase water efficiency. The most common approaches were as follows:

1. Improvements in the efficiency of irrigation systems through rehabilitation and better management systems for water delivery

2. Adoption of specific irrigation techniques (drip trickle, bubbler, sprinkler)

3. Increases in canal flow capacity to reduce water lost through evaporation

4. Improvements in watercourse efficiency achieved by lining and other methods to reduce losses from infiltration and seepage

5. Training in equipment maintenance

6. Studies of water use.

Success with efficiency improvements in agricultural use has been variable. Of 60 completed projects pursuing such improvements, 48 reported having succeeded in improving water use efficiency to some degree. Further, not all projects that resulted in efficiency improvements led to decreased water consumption.

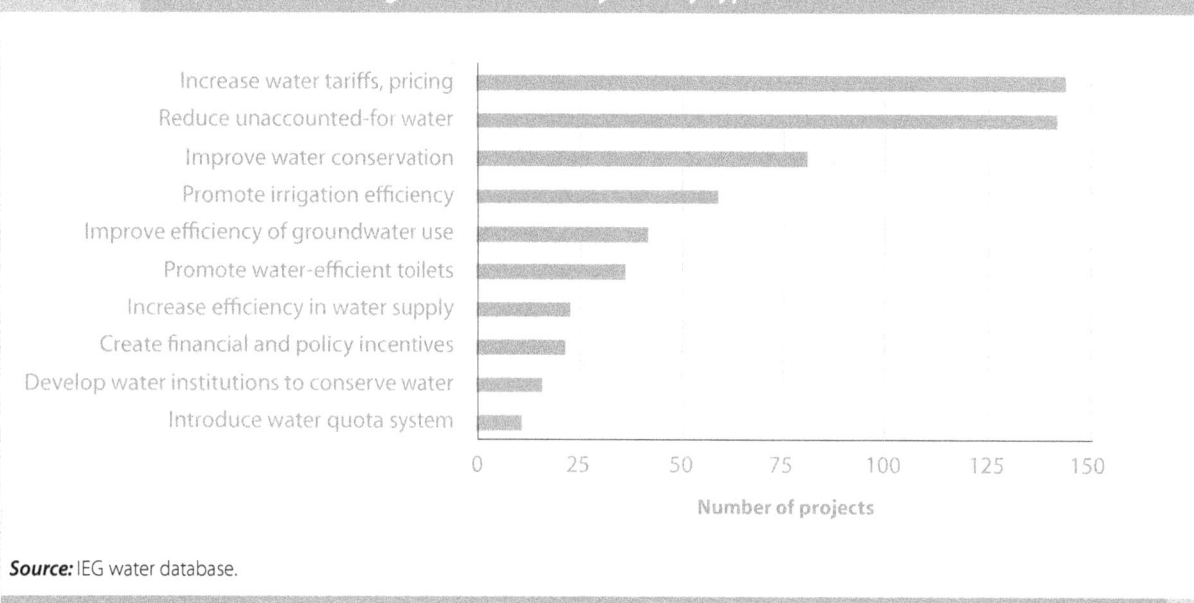

FIGURE 3.7   Water Demand Management Activities by Activity Type

Source: IEG water database.

TABLE 3.1	**Water Efficiency Projects by Subsector**	
Subsector	Number of projects	Percentage of projects that promoted water use efficiency
Water supply	147	27.3
Irrigation and drainage	96	18
Water management and flood protection	35	6.5
Sewerage	30	5.6
Central government administration	28	5.2

Source: IEG water database.

Note: Only the most frequently occurring categories are listed. The total number of DSM projects was 539.

A positive example is the North China Plain Water Conservation Project (P056516). This project combined investments to improve irrigation infrastructure with agricultural support services and forestry and environmental monitoring of the project's impacts on soil and water. Its efforts supported integrated improvements to over 100,000 hectares of irrigated land worked by 257,000 farm households in the provinces of Hebei and Liaoning and in the municipalities of Beijing and Qingdao. The value of agricultural production per unit of water consumed increased by 60 to 80 percent, and nonbeneficial water use was reduced by a sixth. Groundwater overdraft was reduced by 30 percent.

Figure 3.8 shows the distribution of success rates across major areas of intervention. Increasing canal flow capacity was the most successful intervention. The adoption of irrigation technologies with improved efficiency of water use has not necessarily resulted in reduced water consumption. The literature suggests that technologies that increase water use efficiency often lead to increases in planted areas, ultimately increasing consumption (Scheierling, Young, and Cardon 2006).

Where efforts to improve efficiency were tied to financial charges as an incentive to conserve water, the fees were not always collected and the desired conservation was often not achieved.

Reducing Unaccounted-for Water

About half of projects that addressed unaccounted-for water succeeded in reducing it by at least 1 percent.

Within the water supply subsector, reducing UfW was the main activity directed at improving water efficiency. UfW is defined as the share of non-revenue-generating water in total water production. Non-revenue-generating water has two components: it can be real, that is, physically lost by the system, or it can be apparent, indicative of a commercial failure such as inaccurate billing. Addressing UfW mostly involves installing new piped systems or fixing leaks and breaks. The review of the evaluation databases found that of the 1,864 projects that dealt with water, 141 projects addressed UfW. Results of these projects were mixed. Out of the 103 closed projects that intended to reduce UfW, 55 actually managed to reduce it by at least 1 percent. A common problem was that attempts to improve service often involved increases in water pressure, which, absent a thorough leak control program, resulted in increased losses.

Economic Analysis

Of the 539 projects that dealt with water efficiency activities, the evaluation reviewed the economic analysis undertaken by the 373 completed projects.[7] It found that within

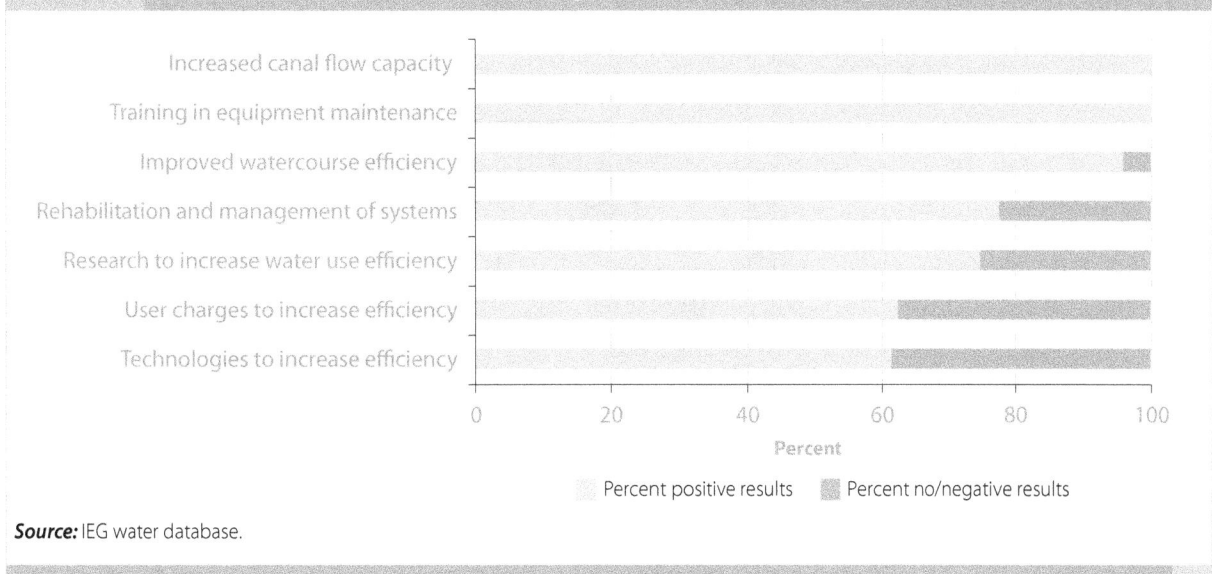

FIGURE 3.8 Success Rates for Demand-Side Management Interventions in Agricultural Water Use

Source: IEG water database.

this subset, economic rates of return (ERRs) were estimated during project appraisal for fewer than half (179). Of these, 136 also provided an ERR at completion; the remaining 43 projects did not. Eight projects calculated ERRs at project completion even though they had not done so at appraisal.

Of the 136 projects that calculated ERRs at appraisal and completion, 59 achieved or exceeded the ERR target at completion.

Of the 136 projects that provided ERR calculations at both appraisal and completion, 59 achieved or exceeded the ERR target at completion. The remaining 77 projects (about 57 percent) did not attain their expected ERRs, at least partly because they did not fully attain the anticipated efficiency gains (figure 3.9).[8]

Using Water Prices to Manage Demand

Until recently, cost recovery has been a core element of the Bank's strategy toward water sector projects.[9] In the WSS sector, the Bank advocated full cost recovery on the basis that covering just the O&M costs has no economic basis and in the long term would result in insufficient investment in the sector. The same pattern holds in the agriculture sector (discussed in chapter 5) (World Bank 2003b, p. 3).

A review of Bank projects that dealt with cost recovery in WSS found a low success rate. The group of projects consists of 133 projects (93 completed, 40 ongoing). The most common strategy used to foster cost recovery is increasing tariffs or water charges (89 percent of the portfolio). Only 15 percent of projects that attempted any cost recovery actually achieved what they set out to do, and only 9 percent of projects that attempted full cost recovery were successful. A subset of 17 completed projects (19 percent) explicitly modified the cost recovery approach during implementation, most frequently opting for a less ambitious approach (less than full recovery) or even dropping the attempt to recover costs altogether. For the projects that achieved full cost recovery, the factor that contributed most to success was improving collection. Most often this involved increasing the capacity and willingness of water institutions to collect fees from beneficiaries. Also, when borrowers were successful in increasing water tariffs as planned, it had a discernible impact on overall project results.

Only 15 percent of projects that attempted cost recovery achieved their targets.

Achieving full cost recovery is challenging for many reasons, which differ sharply between urban and rural sectors.

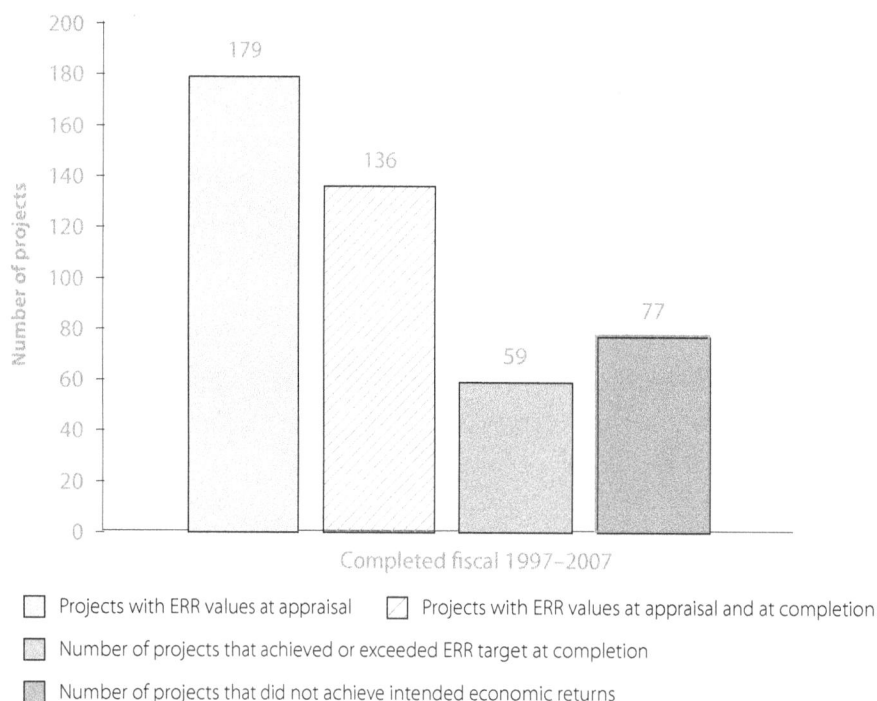

FIGURE 3.9 Completed Water-Efficiency Projects Reporting an Economic Rate of Return

Legend:
- Projects with ERR values at appraisal
- Projects with ERR values at appraisal and at completion
- Number of projects that achieved or exceeded ERR target at completion
- Number of projects that did not achieve intended economic returns

Source: IEG data.

Note: The total number of completed water-efficiency–related projects was 373.

Although it is often easier to levy and collect tariffs for water in urban areas, rapid population growth often keeps the unserved population steady and leads to illegal connections. Further complicating the task, interventions need to deal with the challenges of informal settlements and slums, which may lack roads, street addresses, and public space where pipes can run. In rural areas, the problems include high costs per capita for building, overseeing, and maintaining systems that serve large, sparsely settled areas. Experience has also taught that some people will not be able to afford the full cost of services and that subsidies (possibly including cross-subsidies of some users by others) are often required to increase the access of the poor to services (box 3.8).

The Bank has shifted its stance on cost recovery and now considers it a long-term goal.

The limited success with cost recovery has raised concerns about sustainability in WSS projects. Of 191 completed interventions, 32 allocated inadequate funds to cover the O&M costs of the facilities, and the long-term sustainability of these systems was deemed unlikely. In some cases, by project closing the borrower had not even designated the agency that was supposed to manage and operate the systems. Project completion reports repeatedly cited low tariffs or the failure to collect user charges as the main reasons for suboptimal system functioning and the principal risks to continued operation. The absence of adequate resources to cover the O&M costs of this infrastructure constructed at such great expense is another issue for the sector and the broader donor community to face.

The limited progress on cost recovery has led the Bank to moderate its approach. A recent policy note that spells out the current approach states that the Bank "supports countries to design and implement tariff levels and structures. This assistance to countries is supported by operational research to test existing and new approaches to pricing and subsidies. The World Bank now considers cost recovery for WSS services as a long term goal, but there is flexibility in determining the period of time to obtain this goal. Revenues recovered from users within the near term are to cover the utility's operations and maintenance costs."[10]

The 2006 IEG evaluation *Water Management in Agriculture* concluded that "cost-recovery targets have been wildly ambitious and unrealistic because of inadequate social assessment" (IEG 2006b, p. xix). It also found that expectations regarding cost recovery following the handover of irrigation infrastructure to user groups were unrealistic. And it rec-

Photo courtesy of Edwin Huffman/World Bank.

ommended that simple cost recovery strategies be broadened into strategies for improving water use efficiency.

The report noted that astonishingly few appraisal documents offered a clearly articulated strategy for water use efficiency, whether through some form of volumetric pricing, an area- or crop-based approach, a lower-cost proxy, or a restricted supply approach, and it pointed out the sharp contrast between the vast literature on water pricing and efficiency and actual practice. The literature finds that irrigation demand is inelastic until prices rise to several multiples of service cost, a level that is politically problematic to introduce (Molle and Berkoff 2007; Scheierling, Young, and Cardon 2006). Therefore the Bank and other donors have started to support quotas that limit (primarily) agricultural water consumption. Out of the 11 projects identified by the current evaluation that dealt with quotas, 10 dealt with irrigation, 1 of which has been completed. In that project, the quotas were effective in reducing water use.

Summary

Demands on the world's limited water resources from development and population pressures are increasing and are likely to accelerate. Hence, it is critically important to strike a balance between making good use of water and ensuring sustainability. The Bank has used a variety of methods to help countries manage water resource use in ways that take into account the available supply and the likely future demands on those resources.

Surface water management is intended to help preserve natural resources, improve soil quality, and provide enough water. The 218 projects with at least one watershed management activity performed better than the Bank average on measures of outcomes, sustainability, and institutional development. Projects that used a livelihood-focused approach to watershed management outperformed those that used other approaches, but are considered too costly. However, the full environmental benefits of this approach are seldom accounted for because of lack of data.

Groundwater is increasingly threatened by overexploitation, inadequate environmental flows, and contamination, but the extent of groundwater depletion is poorly understood because data are rarely collected or shared. The number of Bank-financed projects dealing with groundwater has declined since the late 1990s, as projects have rightly shifted away from groundwater extraction. However, the number of projects focusing on sustainability and groundwater conservation has also declined somewhat since 2001, although the 2003 Water Resources Sector Strategy foresaw an increase in attention to groundwater. Projects addressing environmental and resource protection issues that are critically important to the safe, longer-term use of ground-

water resources have tended to perform less well than the extractive projects.

Several countries have established RBOs with Bank support. Efforts to strengthen existing RBOs have had limited success, and the sustainability and effectiveness of newly established RBOs are often in question upon project closure. Basin management often encounters resistance from groups slated to lose power or privileges. In consequence, projects and policies related to basin management often find it difficult to fully achieve their potential benefits.

A relatively small share of Bank-supported projects have helped countries establish systems to collect, analyze, and use hydrological and meteorological data. When monitoring was undertaken, it was often supply driven. More than two-fifths of projects that included hydrological or meteorological monitoring did not tailor systems to meet beneficiary needs, and most of the rest failed to address the issue of who was to use the data. On the positive side, over half of the projects that used monitoring data for disaster prevention and mitigation succeeded in getting the information into the hands of people whose job involved mitigating natural disasters and reducing damage.

Effective demand management is one of several critical approaches in the face of increasing water scarcity. Efforts to improve the efficiency of water use and limit demand in the agriculture sector, the largest consumer of water, have had limited success. Efficiency-enhancing technologies alone do not necessarily reduce on-farm water use. Efforts to modulate demand through water charges have encountered limited success; fixing and enforcing quotas for water use is relatively recent and deserves careful evaluation once more projects have closed. In the water supply subsector, reducing UfW was the main activity directed at improving water efficiency. About half of the projects that attempted to address UfW managed to reduce it by at least 1 percent.

Cost recovery in Bank-supported projects has proved challenging. The successful projects have generally improved the efficiency of water institutions in collecting fees. Limited success has caused the Bank to moderate its approach, but the question of who will pay for uncovered costs remains to be resolved.

ERRs are not routinely calculated even for projects that deal with water efficiency activities. Only about half of projects that aimed at improving efficiency and calculated ERRs upon completion attained their expected returns, at least in part because they did not fully achieve the anticipated efficiency gains.

Chapter 4

EVALUATION HIGHLIGHTS

- The focus on infrastructure rehabilitation in flood-related projects suggests that much flood management infrastructure is not holding up during floods.

- Environmental flow assessments can have significant benefits but still constitute a relatively small percentage of the water portfolio.

- Eighty percent of Bank borrowers have received water quality management support, but few projects measure water quality.

- River and lake projects generally restore physical assets but rarely deal with water conservation or quality improvement.

- Although migration to coastal areas is increasing, annual Bank commitments for coastal and marine management projects have been declining.

- Attention to wetlands in Bank projects has increased since the late 1990s.

Water and Environment

This chapter reviews Bank-supported efforts to address floods and droughts and to improve water quality or increase environmental sustainability, particularly around rivers and lakes, wetlands, and coastal zones.

Coping with Water Disasters

In those parts of the world's temperate zones where rain falls year round in amounts that are generally manageable, the relationship between human settlements and water is a relatively easy one. In contrast, settlements where rain falls in short seasons with significant annual variability have a much harder time. The most severe shortages and excesses of water—droughts and floods—are a source of risk and vulnerability, and their most pernicious effects are felt mainly by the poor.

Severe water crises can be regional, national, or local, and with continued climate change they can be expected to continue to escalate and to eventually cover much wider geographic areas. Food shortages, localized restrictions on trade in food, and population displacements due to water crises are already happening. As a group, weather hazards—drought, extreme temperatures, floods, mudslides, waves and surges, and windstorms (commonly referred to as hydrometeorological events)—affect far more people than all other hazards combined (see appendix figures J.13 to J.16). Over the period 1972–2006, more than 5.2 billion people were affected by hydrometeorological disasters, compared with about 11.5 million affected by all other disasters (including earthquakes, insect infestations, volcanic eruptions, and wildfires).[1] Of course, some countries rarely experienced such events, while others were buffeted repeatedly in the same year, leading to double and triple counting.

Flood Management

Floods are often rapid-onset hazards. Although usually triggered by heavy rainfall, their causes also include the silting up of rivers, the reduced absorptive capacity of soil, flawed infrastructure planning, and inadequate maintenance of existing drainage and flood control facilities. Loss of forest cover and reforestation with inappropriate species contribute heavily to soil erosion, the deposition of silt in rivers, and overly rapid water runoff.

Much public infrastructure does not hold up when flooding occurs. As a result, the Bank finances a great deal of post-flooding rehabilitation. A total of 433 flood-related projects were identified in the water database. As a group, the 164

Photo courtesy of Scott Wallace/World Bank.

projects that focused wholly on flooding performed well: outcomes were rated satisfactory or better for 83 percent.

The most common activity financed by the Bank in response to flood disasters was road rehabilitation. Generally, flash floods and slope instability are the main causes of poor road sustainability. In other words, even rains that are not unusually heavy rains can lead to road loss when adequate drainage and sufficiently dimensioned culverts are not present or road alignments have not been designed to provide maximum slope stability. Top Bank borrowers for floods borrow repeatedly: some borrowed more than 20 times over the evaluation period for flooding issues.[2] Project evaluations show that among the contributors to this problem are lack of maintenance, inadequate design, and poorly considered siting.

The strong emphasis on infrastructure reconstruction in flood-related projects suggests that infrastructure is not holding up when flooding occurs.

A first-order cause for concern is the lack of resiliency of flood control investments. A total of 221 projects that responded to flooding built flood control structures—both preventative and curative. These structures regularly suffer significant damage from the very type of disaster they are designed to ameliorate: in 23 percent of projects, structures built by the project to protect against flooding were damaged by flooding (IEG 2007).

The damage almost always was reparable and did not lead to the total loss of the structures in question, but repeated repairs are a costly drain on public finances.

A similar pattern holds for storm drainage. The evaluation database identified a total of 114 Bank-financed projects that supported the construction and maintenance of storm drains. Small to medium-size cities have been the most common beneficiaries (IEG 2007). Storm drains can help lessen urban flooding, but with even seemingly minor imperfections in conception and design they can end up exacerbating the problem.

Although the Bank has been lending for investments that reduce vulnerability to flooding, such investments have mostly focused on the construction of infrastructure—such as flood control structures or storm drains—even when strategic planning or environmental restoration may have been more strategically appropriate (IEG 2007).

More troublesome, when infrastructure has been the chosen solution, as is often the case for flooding, what the borrowers chose to build with Bank financing often failed to make a significant step forward, possibly because the borrowers lacked the institutions necessary to take charge of the maintenance and management that would make the infrastructure sustainable. This common but unsustainable pattern argues for increased consciousness raising to promote investments in noninfrastructural protective solutions. These can consist of technological solutions, the restoration of natural protections, or the relocation of settlements and facilities to less risky areas (see box 4.1).

Bank projects that aim to reduce vulnerability often focus mostly on infrastructure reconstruction, even when environmental restoration may be more strategically appropriate.

Drought Management

Intense and persistent droughts can have serious economic, environmental, and social impacts (Gleick and others 2009, p. 93). They can cause heavy crop damage and livestock losses, disrupt energy production,[3] and hurt ecosystems across whole regions or countries. The results can include famine, loss of life, mass migration, and conflict, wiping out development gains and accumulated wealth, especially among the poorest (United Nations 2008a, p. 32). Persistent drought can lead to devastating and difficult-to-reverse desertification that threatens agriculture and the climatic and biological diversity of ecosystems. It has been predicted that worldwide, as a result of prolonged drought, about 12 million hectares of land now used for cultivation will be permanently lost to agriculture, threatening the livelihoods of over 1.2 billion people. Moreover, the land area of the planet given over to desert climates is expected to increase by more than 15 percent over the coming years.[4]

BOX 4.1

NONSTRUCTURAL FLOOD CONTROL YIELDS HIGH RETURNS IN THE YANGTZE BASIN

In the Yangtze Basin Water Resources Project in China (P003596, closed in 2003), several grants were provided to finance risk assessments and a support system to improve decision making for Yangtze River flood operations. A flood risk model was also developed by local staff and international consultants to predict flood levels under the project.

The first real test of the Yangtze flood model occurred during the major floods of 1996. As it turned out, the model accurately predicted flood levels, and the decisions based on this model averted flood damage estimated at $15 million. During floods in 1998 and 1999, the system allowed decision makers to avert millions of dollars more in flood damage. According to people involved in the project, the expenditure of $5 million on nonstructural flood prevention brought a return estimated at more than tenfold and led to an important technology transfer.

Source: ICR for the China—Yangtze Basin Water Resources Project (P003596).

Like the occurrence of droughts themselves, Bank lending for droughts has been cyclical (figure 4.1). A flatter pattern would indicate a more proactive stance on both the Bank and the borrower side. (The high number of loan approvals in the late 1990s was due to the El Niño-caused droughts of that period.) That there should be a flurry of activity every time a major drought occurs needs to be seriously reconsidered, given the increasing length and severity of drought events in many countries (box 4.2).

Bank lending for droughts tends to be reactive rather than proactive, and projects in drought-prone areas often do not anticipate drought risk.

Three main subgroups of projects deal with drought. The first consists of projects that react to drought in emergency situations (21 projects). The second consists of projects that seek to mitigate chronic droughts and plan to reduce the financial and social impacts of foreseeable as opposed to ongoing events (89 projects; see box 4.2 for an example). Within the first two subgroups, Sub-Saharan Africa had the largest number of projects (49) and South Asia the highest total loan commitments (over $3.5 billion). The third subgroup consists of drought-affected projects: those that, despite implementing a broad range of interventions, failed to proactively identify drought as a risk to implementation and were forced to conclude in their completion report that drought had an adverse affect on the attainment of project goals (84 projects; these are not included in figure 4.1). Many irrigation projects deal with chronic water scarcity rather than cyclical drought, and are therefore not included in this number.

Drought projects tended to be less successful than flood-related projects, partly because drought mitigation involves substantial behavioral change.

Drought projects were less successful on average at attaining their objectives than flood-related projects, in part because there was little or no rebuilding, but also because mitigation involves a great degree of behavioral change. There were 50 dedicated, drought-focused projects, 72 percent of which achieved a rating of satisfactory.

The successful drought projects relied on very straightforward activities to achieve their objectives—essentially components for which success could be declared as long as disbursement took place. These included purchasing equipment and underwriting the cost of publications development. Activities that achieved 20 percent or less of appraisal expectations tended to be complex and to focus more on results than on outputs. Examples include improving agricultural practices, promoting technologies for more efficient water use in irrigation and water supply, and water conservation (appendix table J.21).

A common lesson from project self-evaluations (Implementation Completion Reports, or ICRs) and independent evaluations (Project Performance Assessment Reports) was that it is impossible to transition to improved and more water-efficient farming technologies under emergency drought conditions, and success is further constrained by the discontinuity of outreach and other activities during the project cycle. Other issues that challenged drought projects included the following:

FIGURE 4.1 Loan Commitments to Drought-Related Projects by Approval Year

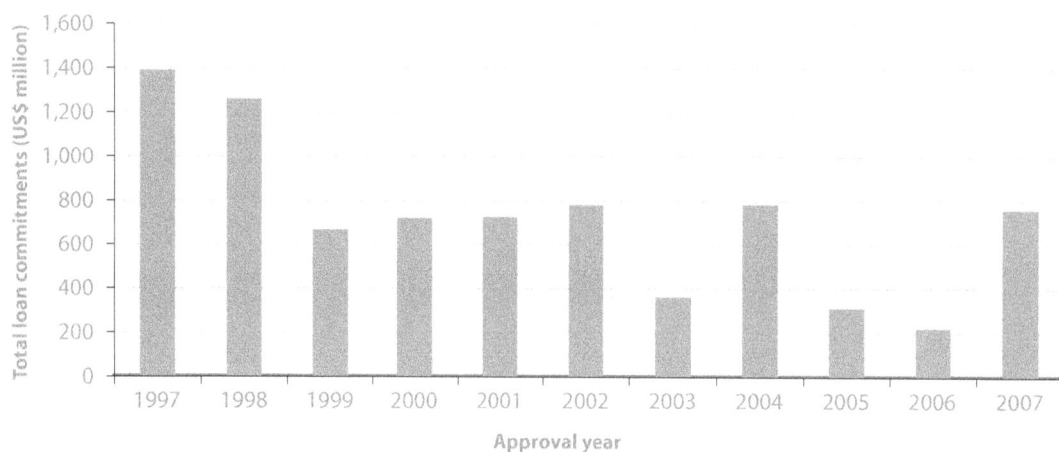

Source: IEG water database.

- Getting the support of stakeholders

- Disseminating new or improved agricultural technologies (drought mitigation)

- Cost recovery for agricultural water.

Completion reports attributed lack of success to:

- Inadequate borrower planning for the impacts of foreseeable events

- Unanticipated difficulties in managing employment creation and income-generating activities

- Failure to monitor and supervise planned activities in a timely manner

- Failure to give due importance to public awareness and community outreach

- Failure to reach or deliver benefits to the most vulnerable target groups.

Experience shows that a transition to more water-efficient farming technology cannot be achieved under emergency drought conditions.

In some circumstances, water insecurity can be managed with the help of dams and reservoirs. The Bank increasingly funds dams that not only serve water security purposes but also provide hydropower, as discussed in chapter 5.

Water and Environmental Sustainability

Most water that is consumed is surface water. Water resources found above ground have both quality and quantity dimensions that are of critical importance to the environment. Addressing environmentally related land use questions when preparing water projects can help to protect water quality in bodies of water still in good condition. In less fortunate areas, improving degraded bodies of water to the point where the water they provide is safe and economical to use requires environmental restoration, an area in which industrialized countries have valuable experience. But even if the quality of water is good, there has to be enough of it to maintain ecosystems and biodiversity. Critical to accomplishing either of these environmental goals is an understanding of how much of the water sitting in a lake or flowing in a river needs to be left there—in other words, how much should be allocated to the natural environment.

Preserving Water Flows for the Environment

According to the Bank's 2003 Water Resources Sector Strategy, the maintenance of water flows essential to the environment needs to be addressed in the design of infrastructure and the recalibration of operating rules in river basins. From 5 to 35 percent of a river's natural flow (depending on the river) can be removed while leaving fragile aquatic environments in good condition (Dyson and others 2003, p. 17). The term "environmental flow" encompasses the flow necessary to sustain an ecosystem and considers the flow's effect on nearby inhabitants and the local economy. The definition states that environmental flow analyses "describe the quantity, timing, and quality of water flows required to sustain freshwater and estuarine ecosystems and the human livelihoods and well-being that depend on these ecosystems."[5]

The Bank's 1993 Water Resources Management Policy Paper signaled a major shift in water sector investments, stating, "The water supply needs of rivers, wetlands, and fisheries will be considered in decisions concerning the operations of reservoirs and the allocation of water" (Hirji and Panella 2003). Until the mid-1990s such assessments were rarely done, either globally or in the Bank's work.

In the evaluation portfolio, 39 projects were identified that at least considered environmental flows. Of those, 18 underwent either a full environmental flow assessment (EFA) or a similar analysis.[6] Attention to environmental flow in ongoing projects rose steadily over a seven-year period (fig-

ure 4.2). To date, over $1.8 billion in Bank-supported projects (1.5 percent of the water portfolio), most of them in the rural sector, have benefited from some kind of EFA. Experience has shown that the reach and potential benefits of EFAs can be substantial (table 4.1 and appendix table J.23).

Water Quality Management

Projects were identified as addressing water quality if they sought to improve water quality parameters, pursued a goal (such as public health or mosquito control) that required adding chemicals to water, or dealt with water pollution likely to be caused by project-funded activities.

The evaluation identified 731 projects—about 40 percent of the water portfolio—that addressed water quality and found that 80 percent of the Bank's borrowers have had projects that fit this loose description (most of them addressing point sources of pollution; box 4.3). They cover a wide range of activities (see appendix table J.24) that reflect the Bank's Water Resources Sector

TABLE 4.1	The Costs and Benefits of Selected Environmental Flow Assessments	
Location	Cost	Benefits
China Hai Basin (ongoing project)	$0.858 million (World Bank), $2.1 million (total)	Determining minimum flows and their scheduling: • Helps ensure that the Bohai Sea, with its globally important ecological resources, will continue to provide significant fishery benefits to China, the Korean peninsula, and Japan • Allows minimum flows to be factored into the planning process • Helps develop priorities for follow-up actions • Helps maintain ecological functions • Helps reduce pollution to preserve environmental uses of water • Helps in the effort to control toxic pollutant loads • Helps avoid overuse of surface water • Aids in the arrest of the decline and deterioration of water resources and damage to freshwater in coastal environments in the Hai Basin • Preserves the Bohai ecosystem and fishery resource • Preserves this seasonal spawning and nursery ground for the larger and more productive Yellow Sea
Lower Mekong Basin	Basin modeling and institutional knowledge base, $9.9 million	• Provides "unifying framework" for assessing ecosystem needs as part of river basin management • Helps avoid changes in flows and salinity from deforestation, dams, and increased abstraction for irrigation • Helps protect the Tonle Sap fishery, which supplies fish to Cambodia and neighboring countries and provides jobs to 1.2 million Cambodians • Helps avoid increased flood frequency and peaks in the rainy season • Helps avoid exacerbated drought conditions, and therefore promotes rice production

Source: "Understanding the Dynamic Links Between Ecosystems and Water in the Mekong Basin" (www-esd.worldbank.org/bnwpp/documents/3/EnvironmentalFlowCaseStudy.pdf.

FIGURE 4.2 Ongoing Environmental Flow Projects by Year

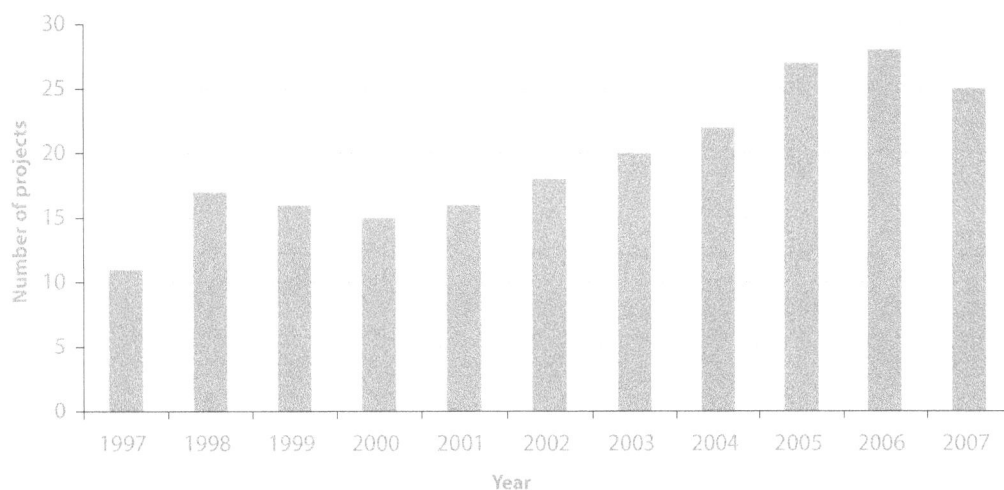

Source: IEG water database.

Strategy. The bulk of the lending (in terms of number of loans) for water quality management has been allocated for technical assistance to strengthen the water resources management capacities of public agencies and to assist them with planning.

Water quality continues to worsen for the top five borrowers for water.

Water quality in the top five borrowers for water continues to worsen, despite the support they have received.[7] The reason is partly that when lending has a water quality focus, the most frequently pursued strategies do not directly confront the physical causes of water pollution—the Bank quite infrequently finances attempts to directly improve water quality within bodies of water.[8] But it also indicates that this problem is far too big for the Bank to solve on its own.

Water Quality Monitoring

Although 40 percent of the water portfolio deals with water quality issues, relatively few ongoing projects actually measure water quality parameters, and the incidence of such monitoring has decreased (figure 4.3). Hence, many projects that should be monitoring water quality are not doing so, and data on water quality produced by Bank-financed projects continue to be in short supply. Furthermore, the quality of data that are collected is often questionable. Although 79 percent of the 61 closed projects that set out to monitor water quality actually did so to some degree, only 66 percent monitored what would be considered technically appropriate parameters. It might have been expected that projects requiring a full environmental assessment would be leading other types of projects in conducting water quality monitoring activities. Yet this proved not to be the case (appendix tables J.27 to J.29 and figure J.18). Recent evaluations by other donor organizations show that they do not necessarily do any better.[9]

Many projects are not monitoring water quality.

Despite the large share of projects dealing with water quality, little improvement can be demonstrated thus far. More than half of projects designed to monitor water quality (see appendix table J.28) cannot show that it is improving, either

BOX 4.3

POINT AND NONPOINT SOURCE POLLUTION

Pollutants make their way into fresh and salty water from point and nonpoint sources. Typical point sources include residential sewerage and industrial end-of-pipe discharges such as effluents leaving wastewater treatment plants. Nonpoint sources are more diffused and harder to identify and therefore more difficult to mitigate. Examples include urban and agricultural runoff and drainage systems. Over two-thirds of Bank projects that deal with water quality management deal with point source rather than nonpoint source pollution, which is common practice, even in highly industrialized countries, because of the difficulty of identifying and mitigating nonpoint pollution.

Source: IEG water database.

FIGURE 4.3 Ongoing Projects That Monitor Water Quality by Year

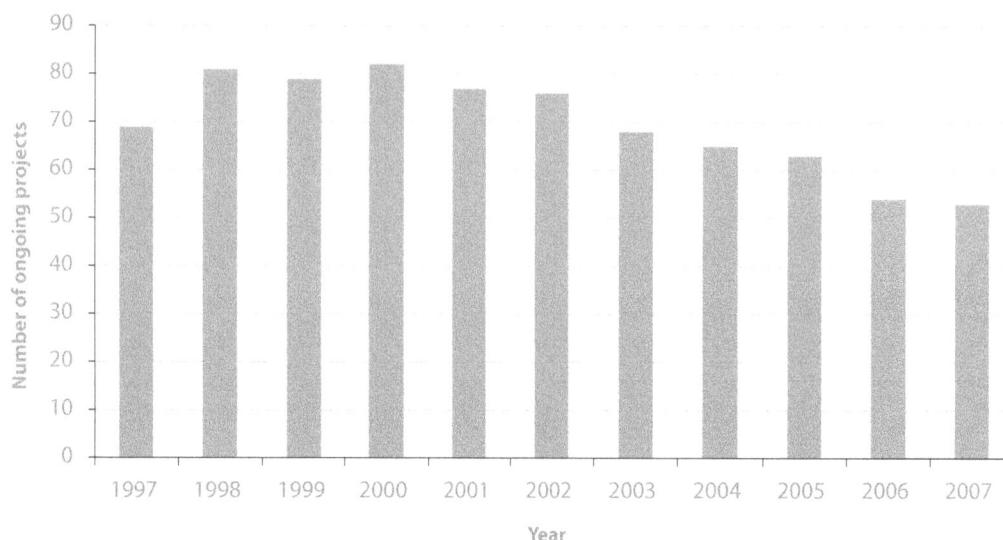

Source: IEG water database.

Note: Over the period as a whole, 102 of 731 water quality management projects monitored water quality.

because it is not, or because water quality data are lacking. Longer-term improvements in water quality are necessary, but achievements will remain hard to identify unless more frequent and thorough monitoring is conducted and results are made transparently available to the general public and the most concerned stakeholders.

Fewer than half of projects designed to monitor water quality can show that quality is improving.

Rivers, Lakes, Coastal Zones, and Wetlands

The majority of the world's population lives near a major water source, be it a river, lake, or coast (UNEP 1999). The quantity and quality of freshwater in these zones are increasingly challenged by growing demands for human use. Wetlands, which occur near many water sources and protect them, have few human uses but important environmental functions, and they are rapidly disappearing. Many bodies of water are shared between countries,[10] making management all the more difficult. The evaluation portfolio contains about 300 projects dealing with water resources management linked to rivers, lakes, and coastal systems.

Rivers and Lakes

Fundamentally, the problems facing rivers and lakes are caused by a lack of adequate attention to the environment. Eutrophication (the presence of excessive nutrients), the presence of invasive species and waterborne disease vectors, changes in land use that alter evapotranspiration rates, sediment flows, and other problems result from human ac-

tivities that do not adequately protect bodies of water. These problems are also a product of lax regulation, insufficient water quality monitoring, and, perhaps most important, a general lack of understanding of the economic importance of clean surface water.

The portfolio review found that the Bank has financed 174 projects dealing with river and lake issues.[11] The majority of these projects (140) were related to rivers only, 30 projects addressed lake-related activities only, and 4 dealt with both. Total (approved) lending associated with these efforts was $14.8 billion. These operations generally financed activities such as stabilizing riverbanks, increasing access to water for economic uses, rehabilitating drainage and wastewater facilities, restoring forests in catchment areas, and managing watersheds (see appendix figures J.20 and J.21 for objectives and activities; watershed and basin management is discussed in chapter 3).

Since fiscal 2000, about a quarter of projects around rivers and lakes have focused on preventing pollution.

Projects' aims and purposes evolved over the period studied. In the mid-1990s, the Bank mainly financed irrigation, aquaculture, industrial use, and water for livestock using fresh surface water. However, population densification and industrial development have increased awareness of the importance of water quality; this is reflected in related lending. Starting from fiscal 2000, 26 percent of all Bank-financed projects around rivers and lakes focused on preventing pollution, mainly from point sources but also from nonpoint

Photo courtesy of Tran Thi Hoa/World Bank.

sources. (For more detail on what these projects did and what worked and what did not, see appendix table J.31.)

Bank-financed projects around rivers and lakes were good at restoring physical assets, particularly flood-damaged infrastructure. But water conservation and water quality improvement were not often attempted, and when they were, the relevant objectives were achieved less than half the time. The lack of success was often due either to overly ambitious objectives or to lack of sufficient data to confirm that the desired changes had taken place.

Projects were good at restoring physical assets, but water conservation and quality improvement were rarely attempted.

An in-depth review of individual completed projects found that those that aimed at controlling excessive water extraction from freshwater bodies have had suboptimal results. Although water conservation is a key factor in managing and protecting rivers and lakes, Bank-supported projects have had little success in conserving water. Nearly all (32 projects, according to self-evaluation reports) attributed the poor results to an absence of financial incentives to conserve water and a failure to meet water pricing targets. Failure to reduce water losses during transportation was the second most common problem in water conservation projects.

Of the 102 completed river and lake projects, 35 addressed the supply of water for economic uses. In 14 completed projects, water availability was increased and access to water supply was improved. Despite efforts to increase the availability of irrigation water from rivers and lakes, 9 projects still reported that appraisal targets were not met (box 4.4).

Coastal Zones

Migration from inland areas to the coast is increasing worldwide, and the United Nations projects that by 2030 about 75 percent of humanity will reside in coastal areas (UNEP 1999). Coastal zones throughout the world have historically been heavily exploited because of their abundance of natural resources.

Between 1997 and 2007 the Bank provided $1.6 billion for 121 water projects that included some coastal zone management activities. (This figure excludes projects that did not explicitly address coastal zone management but did have an impact on coastal zones.) The majority (91 projects) were totally focused on coastal zone management. However, the number of project approvals and annual Bank commitments for coastal and marine management projects declined significantly over the period evaluated, and this type of project does not figure largely among ongoing projects (see appendix figure J.23).

Annual Bank commitments for coastal and marine management projects have been on the decline.

BOX 4.4

HOW THE FAILURE OF IRRIGATION SYSTEM REHABILITATION LED CHINESE FARMERS TO IRRIGATE WITH GROUNDWATER

In the 1970s three large pumping stations, Jiamakou, Xiaofan, and Zuncun, were built to lift water from the Yellow River to irrigate some 123,000 hectares of farmland. The shifting of the river channel and high sediment content caused serious disruptions in the water supply from these stations. At times, access to the river water was completely cut off, and farmers had to rely entirely on groundwater, and this intense use resulted in its overexploitation. Under the Shanxi Poverty Alleviation Project, improved irrigation was to cover 123,000 hectares, and it was expected that this would reduce pressure on groundwater by providing a reliable supply of Yellow River water to the existing Yuncheng Irrigation System. The project aimed to achieve this goal by lining irrigation canals with concrete, rehabilitating water distribution systems, and constructing large-scale sediment traps. By project closing, only 24,400 hectares were being supplied with irrigation water, and most of the target areas continued to rely on tube wells, leading to continuous overexploitation of groundwater.

Source: ICR for the China—Shanxi Poverty Alleviation Project, closed in fiscal 2004 (P003649).

THE BENEFITS OF PARTIAL RESTORATION OF MANGROVE FORESTS IN VIETNAM

Mangrove forests support fisheries by providing breeding, feeding, and nursery grounds for commercially important fish and shellfish, and they lessen the impact of toxic substances on water and soil by serving as natural filters. They also serve as buffer zones against typhoon and flood damage to inhabited areas and limit the intrusion of salinity. But with the expansion of shrimp farming these forests are rarely left unmolested, and to the people living near them they are a source of no-cost wood products, food, and even roofing thatch and traditional medicines.

In Vietnam, coastal inhabitants' disregard of the need to preserve the area's mangroves led to their widespread eradication, with catastrophic environmental impacts. The Vietnam Coastal Wetlands Protection Project tried to balance environmental protection of the mangrove forests with the livelihood needs of people dependent on natural resources. The project was designed to restore wetland ecosystems. Staff highlighted the need to implement reforestation and forest protection while addressing the underlying social causes of encroachment. It was understood that better protection of these areas would inevitably increase coastal and marine productivity in areas close to shore where the smaller boats used by the poor could safely go. Improving the income status of the adjoining communities required a combination of extension, credit, and social support. Contractual measures were created to provide the incentives necessary to ensure that environmental protection would take place. In total, 370 million trees have been replanted along 460 kilometers of coast. By project closing, because of reforestation efforts, the coastal erosion area had been substantially reduced and new land had begun to accrete along the coast: erosion was reduced by as much as 40 percent, and the area of coastline accretion increased by 20 percent.

Source: IEG Vietnam case study research.

The types of activities financed by the Bank included reduction of land-based pollution, management of coastal wetlands, construction of sea defenses to prevent flooding, and development of sustainable coastal tourism. Marine activities included preventing ship-based pollution, conserving marine ecosystems (protecting coral reefs, sea grass, sea turtles, and mangrove trees), promoting aquatic recreation, and developing ports and marinas. The successful partial restoration of mangrove forests, which provide a variety of economic and environmental benefits, is an example of the large benefits that can be had from small-scale efforts (box 4.5).

The Bank's revealed strategic approach to coastal and marine issues is changing. In projects that have closed, the Bank's support focused primarily on integrated coastal zone management plans, the development of policies and regulations, and the prevention of oil discharges and other water-polluting emissions from oceangoing vessels. In contrast, more recently approved (and ongoing) projects focus mainly on strengthening institutions (such as marine and coastal management units) and mobilizing community participation, and on controlling land-based pollution (see appendix figure J.23).

> The Bank's revealed strategic approach to coastal and marine issues is changing to a focus on strengthening institutions and mobilizing community participation.

The establishment of marine protected areas (MPAs) is among the most common approaches to protecting and maintaining marine biodiversity. A recent Bank study found that although less than 1 percent of the area of the world's oceans is covered by some form of protected area status, most protected areas are still not managed effectively (World Bank 2006a). Of the 21 marine ecosystem conservation projects covered in the evaluation, 9 successfully identified or established MPAs: in the Arab Republic of Egypt, Georgia, Ghana, Madagascar, the Philippines (2 projects), Samoa, Ukraine, and Vietnam.[12]

Because the intrusion of salt water into freshwater aquifers is becoming a more serious problem in many coastal areas (whether because of natural processes or because of human activities; see chapter 3), the Bank is financing a number of attempts to combat the problem. A total of 34 projects in the water portfolio undertook activities to prevent seawater intrusion, 19 of which related to the protection of aquifers. Twenty-three projects used nonstructural mitigation measures, such as studies, aquifer monitoring, and water quality testing, and 11 used structural measures to prevent seawater from entering irrigation canals or wetlands. The most common activity in this subset of projects was the construction of barriers to keep seawater from flowing inland. In Turkey, the Cesme-Alacati Water Supply and Sewerage Project (P008985, approved in 1998) made significant achievements in protecting the Ildir freshwater aquifer from seawater intrusion. The $13.1 million project constructed a 700-meter-long barrier—an example of how a low-cost project can have far-reaching effects. However, the general lack of groundwater-quality monitoring, as well as the failure to monitor the impacts of

FIGURE 4.4 Projects Addressing Wetlands and Mangroves by Year

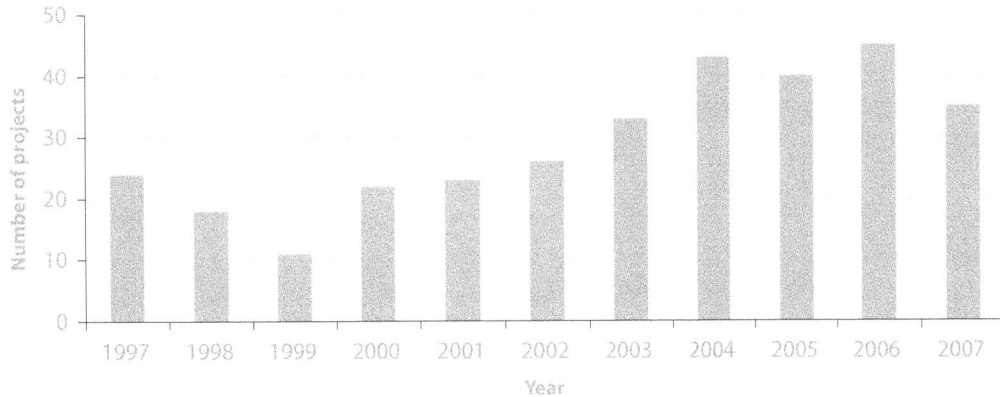

Source: IEG (2007).

the constructed infrastructure on saline intrusion in many projects of this type, makes it impossible to evaluate the results of most efforts.

The Bank has had some success with projects addressing saltwater intrusion.

Wetlands

Bogs, swamps, and marshes filter out waterborne impurities, absorb silt, regulate the flow of water, and add moisture to the atmosphere. Without wetlands, rivers flow too fast, lakes experience destructive algae blooms, and coastlines can be eroded. A little-appreciated facet of wetlands—and a classic example of the tragedy of the commons—is how effective they are at removing pollutants from wastewater. Natural wetlands can remove even heavy metals, sewage, and slaughterhouse waste (IUCN 2009). Unfortunately, half of the world's inland wetlands (excluding large lakes) were lost in the last century. Constructed wetlands, an emerging technology, can be designed to emulate these features, but the Bank

has yet to adopt this approach, and there is quite limited experience for the development community to draw upon.

Attention to wetlands has been increasing in projects since 1997, in part because of the adoption of safeguard policies.

Within Bank-supported projects, attention to wetlands has increased steadily since 1997 (figure 4.4). The increased focus is probably due to the adoption of environmental safeguard policies OP 4.00 (1989) and OP 4.01 (1991, on environmental assessment), which stipulated the identification of potential impacts on wetlands during the environmental screening process.

Evaluating the Bank's work with wetlands is challenged by the imbalance between strong upstream attention and the lack of downstream reporting. When projects close, self-evaluation reports tend to be silent about project achievements in this area, and they rarely pay any attention to the project's negative or unanticipated effects on these fragile

---

BOX 4.6

SUCCESSFUL WETLAND RESTORATION CREATED A REPLICABLE MODEL FOR UZBEKISTAN

As part of the Aral Sea Basin Program, the Aral Sea Regional Water and Environmental Management Project (covering Kazakhstan, the Kyrgyz Republic, Tajikistan, Turkmenistan, and Uzbekistan) successfully restored the Sudoche wetlands, adjacent to the sea, which had dried up as a result of poor water management. The project supported the construction of infrastructure—a dike, an outlet regulator to control water levels, and two new collector channels—to compensate for reduced water flows. The return of water to the wetlands led to the return of various species, and eventually the former inhabitants who had abandoned the zone because of the worsened environmental conditions returned as well. An IEG mission visit in 2007 found newly constructed houses being built right at the edge of the wetlands. The project's approach, designing structures to compensate for reduced flows, was replicated in additional sites by the government of Uzbekistan with its own resources.

Source: IEG Aral Sea case study research.

ecosystems. More than 450 projects were found to address either wetlands or mangrove forests, or both, at the appraisal stage.[13] By project closing, only 98 actually provided information about progress, protection, or the lack thereof regarding wetlands. Of the three completed projects that had wetlands and mangrove restoration as an objective, two achieved satisfactory outcomes.

Summary

The Bank finances a great deal of post-flood rehabilitation, much of it involving infrastructure reconstruction. This strong emphasis suggests that what is being built is not holding up when flooding occurs. The borrowers often do not have the institutions required to take charge of maintenance and management.

Bank lending for droughts tends to be reactive rather than proactive, and projects in drought-prone areas often do not anticipate the risk of drought, even though drought hot spots are well known. Drought projects, because they typically involve behavioral changes, are often less successful than flooding projects at achieving their objectives. Successful projects have generally relied on activities that are relatively easier to accomplish. Projects that focused on such critical issues as improving agricultural practices, promoting technologies for efficient water use, and water conservation often failed to meet their appraisal targets, yet these critical interventions will need to become more prominent.

Information about water quality and environmental flows is scarce, and only a small minority of projects attempt to fill this gap. Lack of data prevents effective planning to address environmental issues related to water quality and availability. Projects designed to monitor water quality cannot demonstrate that it is improving, either because it is not, or because data are lacking. Despite the support they have received, water quality in the top five borrowers continues to get worse. One reason is that when lending has a water quality focus, the most frequently pursued strategies do not directly confront the physical causes of water pollution. Analytical approaches, such as environmental flow assessments, have the potential to integrate environmental concerns into water resources management agendas.

The problems facing rivers and lakes are the result of human activities that do not adequately protect these bodies of water. Fewer than half of Bank projects that attempted such protection achieved the desired result. Controlling excessive water extraction proved particularly challenging. Nearly all of the relevant projects attributed poor results to the absence of financial incentives to conserve water and the failure to meet water pricing targets.

The Bank's lending for coastal and marine management projects has been on the decline. The focus has shifted away from the integrated coastal management plans, development of policies and plans, and pollution prevention that typified the earlier projects that have now closed. More-recent projects tend to focus on strengthening institutions, mobilizing community participation, and dealing with land-based pollution. A number of marine protected areas have been established with Bank support, but these are not managed effectively.

Half of the world's inland wetlands were lost in the last century for lack of protective measures or their enforcement. Within Bank projects, attention to wetlands has increased steadily, but little is known about the outcomes of Bank activities in this area, because data collection and monitoring have been inadequate.

Chapter 5

EVALUATION HIGHLIGHTS

- The use of water for irrigation has increased dramatically in the developing world, but attention to agricultural water resources management and environmental issues has declined.

- Water supply, sanitation, and sewerage activities have been growing in the portfolio since 2004.

- Water supply projects have performed on par with the Bank's overall portfolio and have substantially contributed to improving access to clean water.

- Institutional weaknesses are a common cause of failure in water supply, sanitation, and sewerage activities.

- The poor often cannot afford sanitation schemes even with subsidies or financial aid.

- Although the Bank's share of financing for dam construction for hydropower is modest, the potential is large in developing countries.

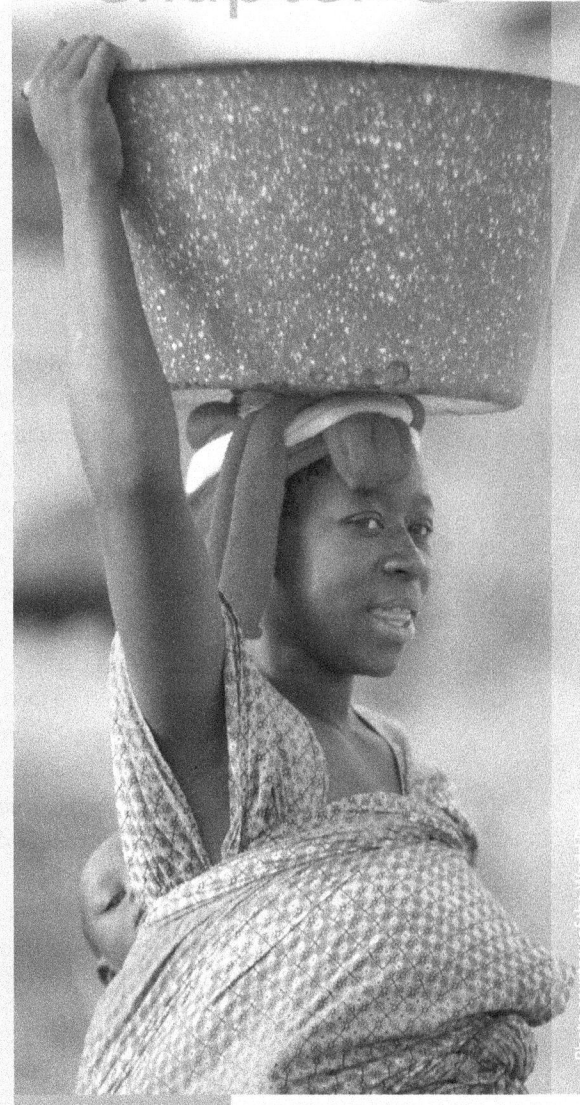

Water Use and Service Delivery

Water has almost as many uses as there are human endeavors, and the delivery of water services, a focus of the MDGs, is important to meeting the Bank's poverty reduction mandate. This chapter summarizes the trends and the findings of the IEG evaluation regarding the delivery of some of the most important water-related services: water for agriculture, water supply, basic sanitation, sewerage and wastewater treatment, and dams for hydropower.

Irrigation

Irrigated land makes up about a fifth of the arable area in developing countries, but it accounts for about 80 percent of all water use. The volume of water used annually for irrigation in the developing world has doubled in the past 40 years. But that increase has slowed because of heavy draws on groundwater aquifers and competition for water from other sectors (World Bank 2006a, p. 3). At the same time, Bank lending for agricultural water management has decreased in every decade since the 1970s (IEG 2006b).

The productivity of irrigated croplands has increased dramatically in the past half century. For example, in that period the production of rice and wheat increased 100 percent and 160 percent, respectively, with no increase in water use per bushel. "However, in many [river] basins, water productivity remains startlingly low," the International Water Management Institute's (IWMI) Comprehensive Assessment notes (Molden 2007). Without continued improvements in agricultural productivity, or major shifts in farming locations, the amount of water required for agriculture will grow by 70 to 90 percent by 2050, according to the IWMI's report.

In 2006 IEG published an evaluation of agricultural water use that reviewed the experience of 161 projects in 56 countries (IEG 2006b). This evaluation draws on that work. The 2006 evaluation found that attention to agricultural water resources management had decreased steadily during the period studied and that the discussion of agricultural water charges had almost disappeared from Bank Country Assistance Strategies (see appendix F). The notable exception to this trend was the Middle East and North Africa, where concerns about agricultural water resources management top the agenda because rapidly growing urban areas are competing with agriculture for scarce water. The study further found that although measures to increase the efficiency and sustainability of irrigation and drainage were among the most important development objectives of agricultural water management projects, attention to environmental issues had steadily fallen since the mid-1990s. The IEG report also concluded that more attention to water rights was important, though progress would be challenging where ownership of land and water assets was unclear.

Attention to agricultural water resources management has declined steadily, as has attention to environmental issues.

This evaluation reviewed the experience with community management in 62 projects that created or supported water user associations (WUAs). It found that about 87 percent supported irrigation and drainage. The remainder were in the WSS sector. The main aims of both clusters of projects were to implement the user-pays principle; to maximize the revenue collected through maintenance of accounts receivable and bill collection; and to reduce the cost of scheme maintenance, often through the use of local labor. About 80 percent of the projects created new WUAs. Albania, Indonesia, the Kyrgyz Republic, Mexico, and Pakistan established federations of WUAs to manage and operate main canals and to coordinate the use of jointly owned large agricultural machinery and water equipment among their members. About six other projects foresaw setting up such federations at an unspecified future date.

WUAs that did not work effectively often had received insufficient training of their members and lacked member ownership.

WUAs were established as planned less than half the time, but three-quarters of those that were established were reported to be working effectively at project closure. Farmers' lack of motivation to organize themselves just to achieve a goal imposed by the government (often in line with the donor's agenda), was the key reason underlying the low rate of WUA establishment. Ineffective WUAs were often those that did not provide sufficient training for members, which led to uncertainty and conflicts regarding their rights and obligations, culminating in a low level of willingness to pay for service. About 15 percent of WUAs failed to achieve long-term sustainability because of their own weak administrative and financial capacities: lack of basic equipment (office space, computers, and telephones) and vehicles was often the cause.

Four main lessons can be drawn from the WUA experience:

- Developing the capacity of WUAs is a long-term process that often cannot be completed within the span of one project.

- Training is integral to WUA development and works best when it entails full participation of members in the things the association is actually responsible for: planning, operation, and maintenance of water systems.

- If the aim is full cost recovery or even just recovery of operation and maintenance costs, water charges need to be realistic; that is, they must be adequate to cover the actual expenses associated with running the systems. In-time water distribution and proper maintenance contribute to a higher ratio of water fees collected to fees assessed.

- Transfer of public water systems to user groups can create a sense of energy and empowerment that can greatly improve system efficiency, household incomes, and enterprise-creating investment. The transfer process will not be successful and sustainable unless the members of the WUAs perceive real benefits in short order.

Water Supply and Sanitation

Even with 87 percent household water service coverage (UNICEF and WHO 2006, p. 23), over 1 billion people worldwide—one-sixth of the earth's population—lack access to safe drinking water. Bank involvement in WSS has been growing

Photo courtesy of Arne Hoel/World Bank.

steadily over the past decade (figure 5.1) and comprises 550 projects (29 percent of the 1,864 projects in the water portfolio). The number of projects with WSS activities has been growing steadily since 2002 (except for a dip in 2006), probably because of an overall increase in infrastructure projects, emphasis on the MDG targets for accessibility to clean water and basic sanitation, and endorsement of the World Bank Water Resources Sector Strategy in 2003. Many of these projects have included social and educational activities as well. Total lending approved between 1997 and 2007 for WSS projects was just over $13 billion.[1]

Water supply, sanitation, and sewerage activities have been growing in the portfolio since 2002.

The evaluation found that within these projects, "hardware" (physical assets) and "software" (institutional and financial capacity building, technical and financial management support, training, and studies) were quite evenly balanced so as to increase the capacity of local institutions to provide additional services without further Bank support. Table 5.1 shows that five of the seven most frequently undertaken activities involved "software"; 57 percent of projects addressed one or more of the above activities.[2] Several methods—including activities in project administration, writing plans and proposals, accounting, computer training, and putting in place participatory structures such as water users groups— were used to build institutional and financial capacity.

The projects generally balanced physical infrastructure with capacity building, technical assistance, and training and studies.

Urban Water Services

The evaluation portfolio contains 556 projects dealing with urban water services, 229 of which focused exclusively on provision of urban water. Water services in urban areas are often subsumed under other rubrics and consist of more than just WSS (see figures 5.2 and 5.3).

On average, urban water projects performed about as well as the overall water portfolio over the evaluation period. However, they started from a lower base (65 percent satisfactory outcomes in 1998), and there has been a steady improvement in their success rate over time (80 percent were rated satisfactory in 2007). Ratings for urban projects vary by Bank Region (figure 5.3). Fourteen countries with multiple urban water projects achieved satisfactory outcomes on all relevant projects (table 5.2).

As a group, the 356 urban water projects in the portfolio have performed about the same as the overall portfolio, with a steadily improving trend.

The highest-rated urban water projects (those with "highly satisfactory" outcomes) successfully completed the construction of physical infrastructure *and* successfully achieved institutional reforms or structured or restructured utilities. The least successful projects, in contrast, did not complete their infrastructure components, usually because the borrower failed to address dysfunctional institutional arrangements. It is therefore a positive sign that institutional development is an increasingly common activity in ongoing projects. Nearly 30 percent of completed urban water projects dealt with the risks to water supply either directly or indirectly. In general, projects that explicitly addressed risk were concerned with preparing for or recovering from

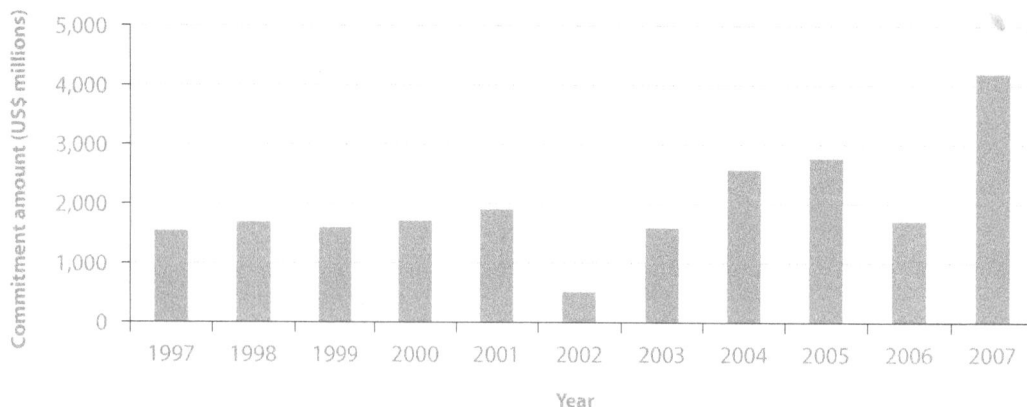

FIGURE 5.1 Bank Commitments for Water Supply, Sanitation, Wastewater Treatment, and Sewerage

Source: IEG water database.

Note: Includes both loans and grants.

TABLE 5.1 The Top 10 Activities for Water Supply and Sanitation Projects		

Activity	Number of projects	Percentage of all projects
Institutional strengthening and capacity building[a]	166	30
Rural water supply and sanitation	149	27
Urban water supply and sanitation	148	27
Technical assistance[a]	110	20
Environmental management[a]	73	13
Studies[a]	69	12.5
Training[a]	69	12.5
Wastewater treatment	66	12
Financial capacity building	49	9
Equipment purchase	47	8.5

Source: IEG water database.

a. Activities that can be categorized as "software."

natural disasters, as well as addressing water quality issues or facing scarcity. These projects performed somewhat below the rest of the urban water portfolio (64 percent had satisfactory outcomes). Another way to help improve the performance of water utilities is through benchmarking, which can drive utilities to improve performance relative to competitors or to utilities in other countries (box 5.1).

The most successful urban water projects both completed physical infrastructure and achieved institutional reforms.

Rural Water Services

The world's 1 billion people without access to clean water live mainly in rural areas.

A major challenge for developing countries in achieving the MDGs, and in particular the MDG target with respect to safe drinking water and basic sanitation, is finding ways to provide sustainable water supply and basic sanitation in small towns and rural areas. As noted earlier, about 1 billion people are without access to clean water, and they are located mainly in rural areas (UNICEF and WHO 2006).

The evaluation portfolio includes 218 projects that had at least one component dealing with rural water supply, representing $13.9 billion in World Bank support. In 96 of these projects, rural water supply was a primary focus. Seventy-two percent of the 96 projects achieved a satisfactory outcome rating. The dedicated projects were rated likely to be sustainable (or better) in 67 percent of cases.

FIGURE 5.2 Completed Project Activities for Urban Water Services

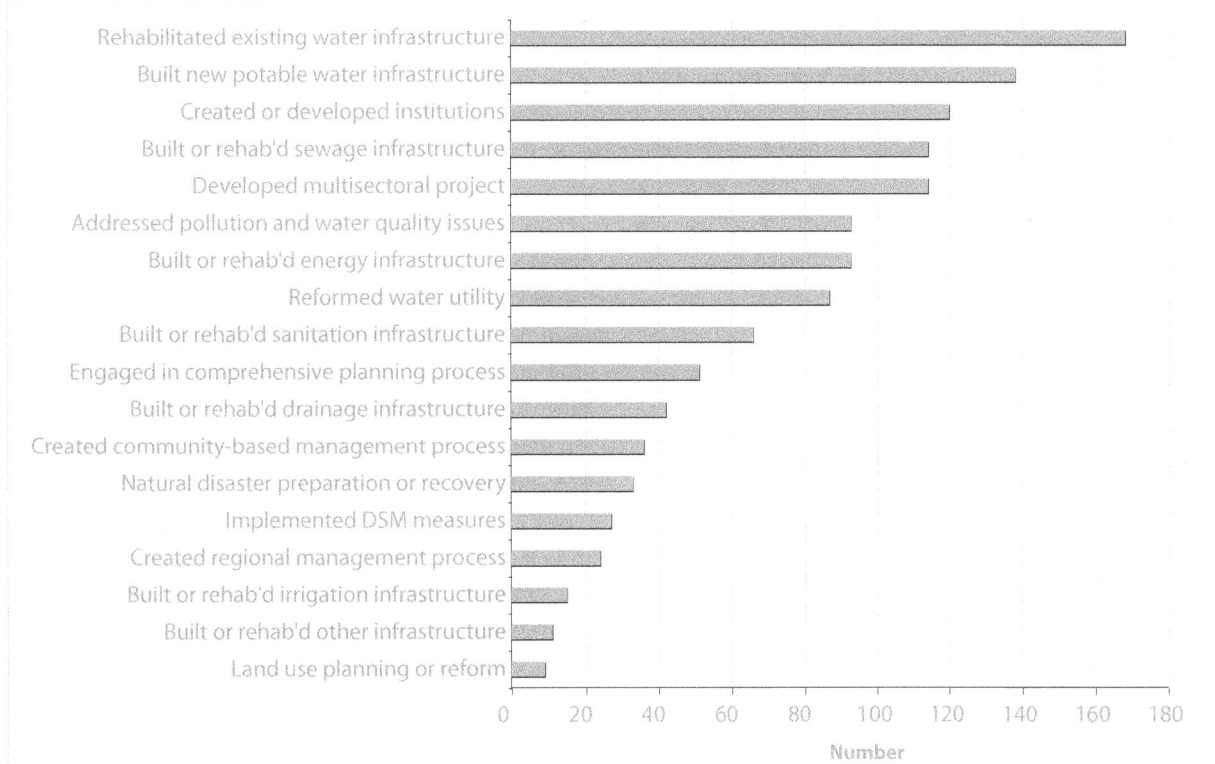

Source: IEG water database.

FIGURE 5.3 Share of Urban Water Projects Rated Satisfactory by Region

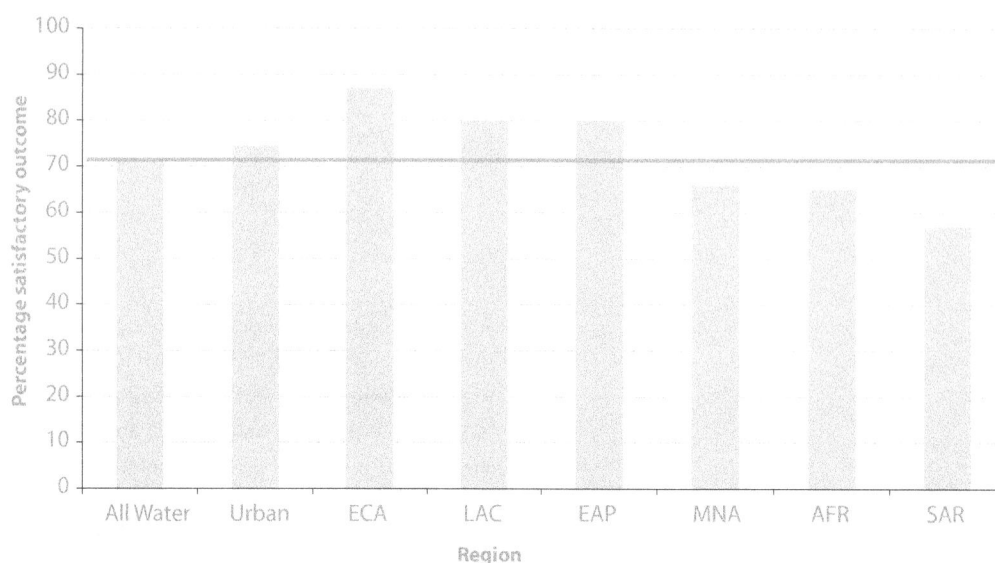

Sources: IEG water database.

Note: Dark horizontal line indicates the average for all urban water projects. AFR = Africa, EAP = East Asia and Pacific, ECA = Europe and Central Asia, LAC = Latin America and Caribbean, MNA = Middle East and North Africa, SAR = South Asia.

Bank-supported rural water supply investments have helped increase rural households' access to safe water. Beneficiary data from 17 completed projects indicate that World Bank financing has made a modest yet important contribution to improving access to rural water supply over the past decade (figure 5.4). However, as the figure also indicates, a more broad-based effort is still needed if the donor community is to provide access to improved water sources in rural areas, although in aggregate terms, progress in China and India alone will be enough to meet the MDG target.

Improved access to rural water supply helped raise the daily water consumption of rural families who previously fell below minimum standards. This improvement, in turn, produced other favorable impacts in health and economic standing. Savings in distance traveled and time spent obtaining water varied significantly by type of water service, with the largest gains accruing to those who used a house connection, followed by those who used a shared pipe connection or a shared source. Women and children were often key beneficiaries of time savings (box 5.2). Reductions in time spent per household in water collection freed up labor for other income-producing activities and schooling. Whether beneficiaries rely solely on the new system for their drinking water greatly influences the impact on health. Reasons for using other, less safe sources include lack of information or training, long distances to safe water sources, and social and economic incentives or disincentives.

Sanitation

Progress on sanitation has been limited: 2.6 billion people—40 percent of the world's population—still lack access to improved sanitation.[3] The transition away from expensive and wasteful waterborne sanitation has been slow—which goes a long way in explaining the persistence of the problem.

TABLE 5.2 Countries with 100 Percent Satisfactory IEG Outcome Ratings for Urban Water Projects

Country	Number of projects
Bolivia	7
Iran, Islamic Rep. of	6
Afghanistan	4
Peru	4
Benin	3
Cambodia	3
Georgia	3
Jordan	3
Korea, Rep.	3
Romania	3
Ukraine	3
Eritrea	2
Estonia	2
Mongolia	2

Source: IEG data.

The review of sanitation-related projects found a preference for capital-intensive works: 312 projects supported wastewater treatment, while 115 projects addressed household sanitation. A number of the latter took a lower-tech approach: 47 projects supported the installation of latrines, and some also constructed shared public toilets. In eight countries alone (Benin, Bolivia, China, Ghana, Pakistan, Panama, Paraguay, and Sri Lanka), the Bank financed a total of 165,297 latrines.[4] Work with waterless sanitation

(composting toilets and dry pit latrines) was undertaken in 23 projects in the water portfolio (appendix table J.34). Seventy-three projects provided household sanitation. According to Bank documents, the Bank is providing the materials for the basic structure—cement, porcelain pans, slabs and toilet seats, and connection and vent pipes. Earlier lessons from experience with the use of expensive latrine models suggested that the use of materials not locally available or too expensive for beneficiaries (or both) limits the uptake

FIGURE 5.4 Share of the Rural Population with Access to an Improved Water Source by Country

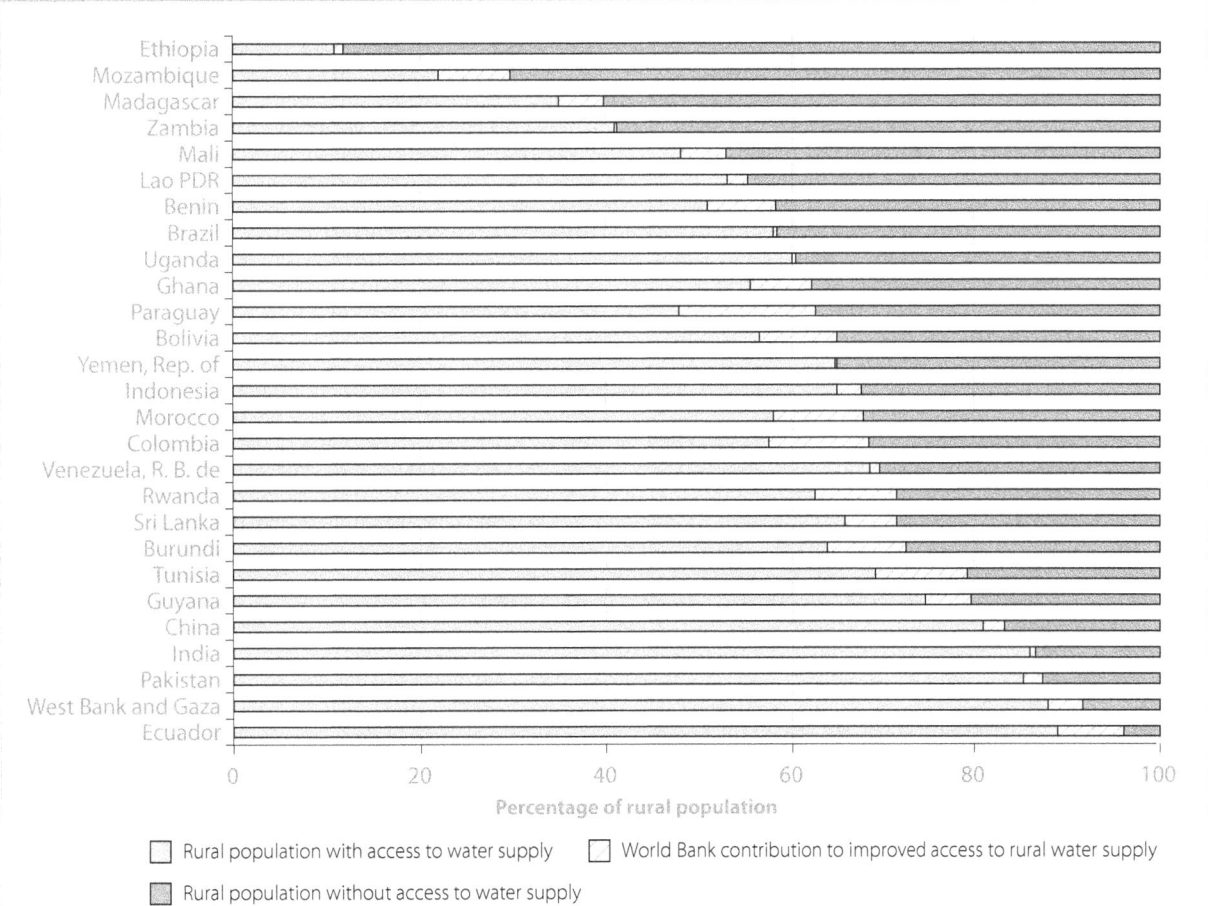

Sources: UNICEF and WHO (2006) and World Bank project self-evaluation reports.

GENDER AND WATER: FINDINGS OF THE IEG GENDER EVALUATION

The 2009 IEG gender evaluation examined the degree to which water services were provided in a gender-sensitive manner in 12 countries. It distinguished the improvements in water service delivery derived by taking women's concerns directly into account. Field observations and desk reviews found that a safe and more convenient water supply had improved the quality of life and productivity of rural women who used project-supplied water in their homes. Easier access to safe water (and an assured supply) allowed women to provide a cleaner home environment with less effort. They washed their own and their children's hands more often, and the improved health of their children allowed the mothers to be away from the home doing productive activities more days per year. Safe water close to the home also helped women keep food preparation areas clean. Clothes washing and cooking were simplified and less time consuming.

The targeted support for women, the evaluation noted, had been at a low level in about half the countries, and sustainability was a concern in some projects.

Source: IEG (2010).

of sanitation schemes and renders them not replicable. In the three most recent projects, the use of locally available materials for latrines was stipulated.

Forty percent of the world's population still lack access to improved sanitation.

Many countries are reluctant to borrow significant amounts of money from the Bank for onsite sanitation, particularly when the materials to be purchased are not capital intensive. Still, there are significant Regional variations, with countries in East Asia and Africa borrowing more for sanitation than countries elsewhere (appendix table J.41).

Subsidies for Basic Sanitation

In order to deliver affordable sanitation services, reduce health risks, and minimize negative impacts on the environment at a reasonable cost, the Bank supports subsidies for basic sanitation infrastructure (for example, latrines and cesspits of various types) required by households and communities for safe management of excreta. The evaluation identified 51 projects that provided a sanitation subsidy, two-thirds of them in rural areas (appendix figure J.26). About three-quarters of the projects provided a partial subsidy (ranging from 20 to 96 percent), which often varied with beneficiary income or the level of service chosen or provided. Almost half of the projects (25 of the 51) required a cash contribution, although in-kind contributions, mainly labor and construction materials, were also accepted in 15 projects. Despite the subsidies, poor households still struggled to meet the requirements. Out of 25 closed projects, 14 (56 percent) provided evidence that the sanitation schemes were not affordable for the poorest beneficiaries.

The poor often cannot afford sanitation schemes, even with subsidies or financial aid.

Over the past 10 years, subsidies have increasingly been directed toward marginalized groups, with beneficiaries selected using characteristics other than income. (Women, children, the elderly, and indigenous groups are all examples that the evaluation came across.) Projects are more likely now to give preference to beneficiaries who are organized. A clear advantage of an organizational structure is that it can act as a liaison between project implementers and beneficiaries, allowing the latter to state their needs and technology preferences, and the former to accelerate and coordinate the generation of a financial contribution. The downside is that the poorest groups tend to have the weakest organizational capacity, and so may lose out.

Subsidies have increasingly been directed toward marginalized groups, but their lack of organizational capacity may still limit access.

The results show that, on average, subsidies help little with the uptake of sanitation. There is no statistically significant relationship between the level of subsidies and the percentage of beneficiaries taking up the improved water or latrine programs (the *t* statistic on the subsidies variable is 0.76; see appendix tables J.35 to J.37).

Sewerage and Wastewater Treatment

Absent adequate wastewater treatment, the pollutants in wastewater can wind up in watercourses, from which they eventually get recycled, with adverse implications for human health and the environment. Of course, in less dense areas, and in dryer zones where streams are scarce, this is much less of a problem. Constructing sewerage removes pollutants from densely populated areas, and wastewater treatment infrastructure helps to prevent untreated wastes from being discharged into fragile ecosystems and protects downstream users.

Effective collection of urban wastewater has become a critical problem in most developing countries. Sewerage and related investments were provided in 312 projects in the evaluation. In the construction of new public sewers and the rehabilitation of existing ones, urban areas predominate—as is to be expected, given that wastewater treatment is not cost-effective in rural areas. About 95 percent of the 312 projects reviewed were implemented in urban areas. Sixteen projects involved activities related to rural sewerage and wastewater treatment.

A comparison of activities of the closed projects with those of ongoing projects reveals how the Bank's approach to wastewater treatment and sewerage has evolved (see appendix figure J.25). In particular, the construction and rehabilitation of wastewater treatment plants have increased enormously, as borrowers faced growing amounts of wastewater generated by rapid population growth and urbanization. Moreover, in times of increasing water scarcity, treatment plants are a popular strategy to increase water use efficiency.

The Bank's approach to wastewater treatment and sewerage has evolved with demand, focusing on construction and rehabilitation of treatment plants.

High-tech treatment facilities that produce extremely good-quality wastewater may be a luxury that only middle-income countries can afford. Hence, a broad approach to sanitation that provides low-tech, step-by-step improvements may be preferable to high-cost treatment plants in countries that have a lot of catching up to do.

Over two-thirds of projects achieved successful results. When project achievements fell short of expectations, institutional weaknesses in the sanitation sector were frequently the cause (box 5.3). IEG also saw a number of treatment plants functioning below design capacity because households were not getting connected to the systems; this is an area where more emphasis is needed.[5]

The Asunción Sewerage Project in Paraguay, described in box 5.3, illustrates this point. The extent to which overcapacity in Bank projects is the result of overdesign of the treatment works is unclear. It could also be the case that the client or the Bank, or both, is overly optimistic about the fraction of the population willing to pay for a house sewerage connection. Although the Bank often talks about the "populations served" by treatment works, the individual household actually receives no direct benefit from treatment works and is often perfectly satisfied with discharging its waste into public drains or a drainage field. The beneficiaries are those downstream of the treatment works, not the polluting households themselves.

SOME REASONS FOR FAILURE OF SEWERAGE WORKS

The appraisal document for the Algeria Water Supply and Sewerage Rehabilitation Project (P004974, $250 million, closed in 2004) proposed the rehabilitation of sewerage networks and sewage treatment plants in several places around the country so that when used water was released back into the environment it would not cause problems for downstream users. In the end, inadequate institutional arrangements prevented the attainment of environmental and public health benefits. There was effectively no responsible authority to oversee the operation, and maintenance of the rehabilitated systems and other investments was dropped. Expected modifications to the tariff structure were not made because the borrower was reluctant to deal with the political risk.

In Nepal's Urban Water Supply and Sanitation Rehabilitation Project (P010370, closed in 2000), the government sought Bank support for rehabilitating two sewage treatment plants and providing 10,000 new sewerage connections in 12 towns. The project failed to achieve most of its objectives, including the sewerage construction and rehabilitation works, largely because the managerial and operational capabilities of national water utilities were weak. As in the Algerian example, tariff increases were inadequate in Nepal because of government hesitancy.

The Asunción Sewerage Project (Paraguay, P007926, closed in 2004) planned to expand sewerage services in that city through the provision of approximately 50,000 new house connections, construction of about 550 kilometers of sewage collectors, installation of 24 kilometers of sewage interceptors, construction of preliminary treatment facilities and a new sewage outfall, and rehabilitation of three existing sewage outfalls serving the city's sewerage system. However, no families in Asunción were connected to the sewerage system as a result of the project, because the local water utilities were incapable of dealing with the complexity of the multiple civil works. The project completion report found that there was little point in constructing expensive facilities without also creating the institutions that can effectively run the systems and ensuring that the target population gets connected.

Sources: Project ICRs and IEG water database.

When projects fell short of expectations, it was usually because of institutional weaknesses.

About a third of water supply projects and fewer than half of sewerage and wastewater projects undertook economic analyses, and a majority of these did not meet their targets.

Analysis of the 312 projects dealing with wastewater treatment and sewerage revealed that the expected number of beneficiaries of sewerage work and the impacts (in terms of both water quality and volume treated) of wastewater treatment consistently fell short of appraisal expectations (figure 5.5). A drop in the number of expected beneficiaries in more recent projects reflects a shift in focus (toward leaving household connections to borrower agencies) and more realistic appraisal targets.

Economic Analysis for Water, Sanitation, Sewerage, and Wastewater Treatment Projects

A review of project completion reports shows that about one-third (35 percent) of completed water supply projects and fewer than half (48 percent) of completed sewerage and wastewater projects undertook an economic analysis at project completion to test the efficiency of the investments. Forty percent of water treatment and sewerage projects and a similar share (38 percent) of water supply projects for which an economic rate of return (ERR) was calculated achieved or exceeded their ERR appraisal targets (figure 5.6). In most cases, ERRs for water treatment were estimated based on the expectation of economic benefits stemming from broader access and availability of treated water, although many of these projects also expected direct economic benefits from the improved sanitation (delivered through construction or rehabilitation of sewer networks) within the target area, irrespective of treatment. Although many completion reports did not provide a justification for not calculating ERRs, about two-fifths of reports for water treatment projects that lacked ERRs argued that quantification of the economic benefits of improved health,[6] improvements to the environment, and better quality of life was too difficult. Even among ongoing water treatment and sewerage projects, only one-third estimated an economic or financial rate of return at appraisal (figure 5.7). Of course, this will make it impossible for any ex post evaluation to compare results at completion with any economic goals or objectives.

Multipurpose Dams and Hydropower

There are now more than 45,000 large dams in 140 countries worldwide.[7] Although dams help make water resources more available in some regions, they are also controversial. Since they modify the water cycle, they can have negative effects on livelihoods, health, and the environment if not planned and managed well. For instance, when sediments no longer reach the coast because they are blocked by dams, the result can be land subsidence in coastal areas. Rivers may run dry if too much water is extracted for irrigation upstream. Lower river flows can also contribute to diseases such as bilharzia and malaria. Populations upstream of a dam can be displaced by the new reservoir, and those downstream that relied on the natural flows for agriculture and fishing can be adversely affected. Nevertheless, given

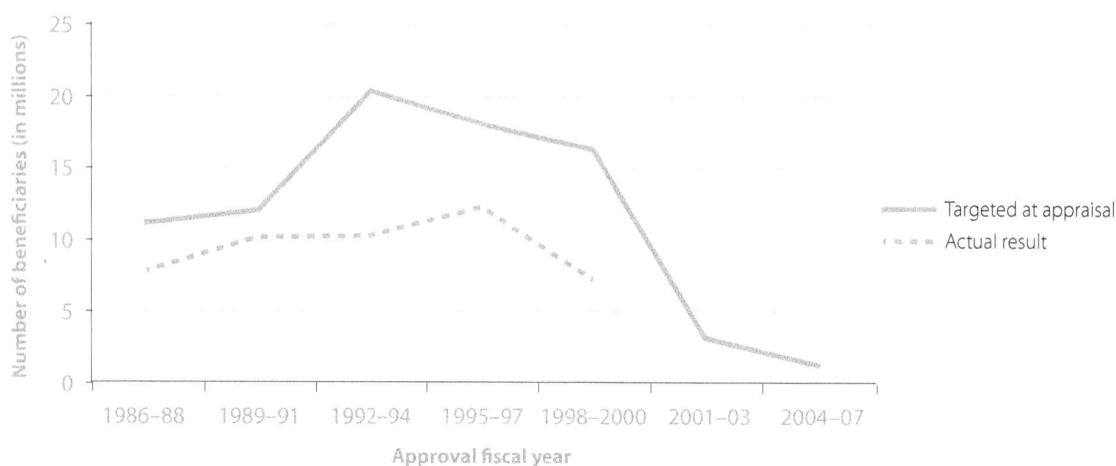

FIGURE 5.5 Expected and Actual Beneficiaries of Sewerage Works

Targeted at appraisal
Actual result

Approval fiscal year

Sources: World Bank project documents.

FIGURE 5.6 Water Supply Projects Reporting Economic Rates of Return

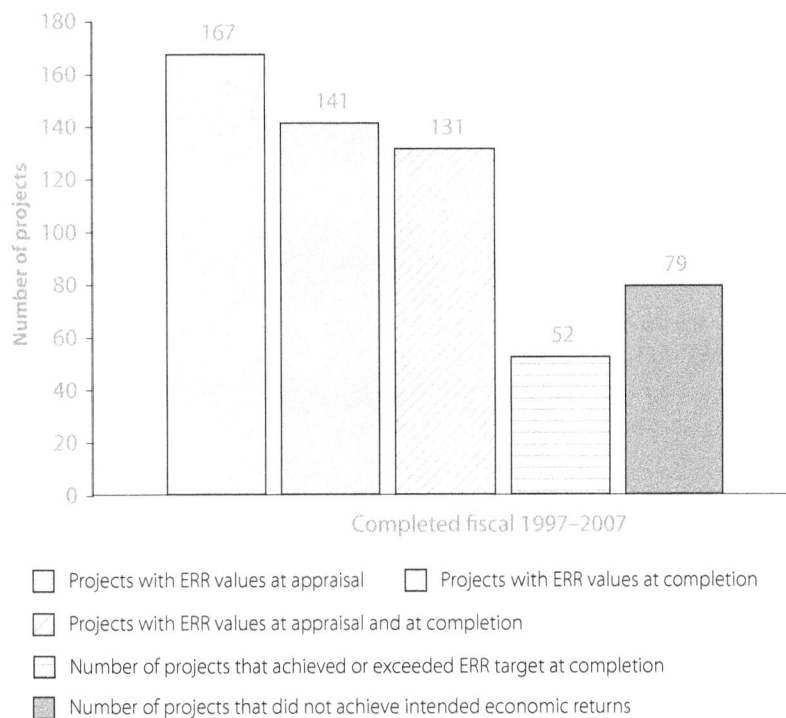

Projects with ERR values at appraisal
Projects with ERR values at completion
Projects with ERR values at appraisal and at completion
Number of projects that achieved or exceeded ERR target at completion
Number of projects that did not achieve intended economic returns

Source: IEG water database.
Note: N = 366.

FIGURE 5.7 Projects Calculating Economic or Financial Rates of Return

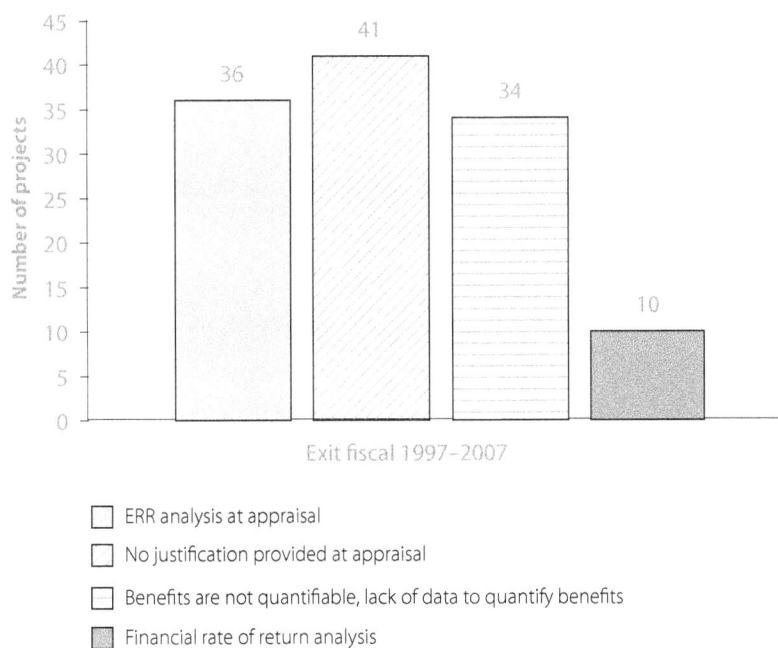

ERR analysis at appraisal
No justification provided at appraisal
Benefits are not quantifiable, lack of data to quantify benefits
Financial rate of return analysis

Sources: World Bank project documents.

the changes being wrought by climate modifications, hydropower's contribution to clean (that is, non-carbon-emitting) energy is gaining increasing importance.

Hydropower is an increasingly important source of clean energy.

After a peak in the mid-1990s, dam construction slowed in the late 1990s following an increasingly difficult dialogue among nongovernmental organizations, the private sector, governments, and international organizations. In 1997 the World Commission on Dams was established to review the development effectiveness of large dams and develop internationally acceptable criteria, guidelines, and standards for large dams. These have since helped guide decisions by the World Bank and others.

The Bank has recently again increased its financing for dam construction, in many cases for multipurpose dams that include the provision of hydropower to help countries develop renewable energy resources (box 5.4). Annual commitments for projects involving dams have recovered from a low of $280 million in 1999 to nearly $700 million in 2007 (appendix H).

The Bank's recent return to financing dams has often included multipurpose dams that also provide hydropower.

Still, the Bank's share of worldwide dam construction for the provision of hydropower is modest, although the potential is large. In developing countries, economically feasible hydropower potential exceeds 1,900 gigawatts (GW), 70 percent of which is not yet exploited. This compares with current installed capacity of 315 GW in Europe and North America and 730 GW worldwide. About 7 percent of economically feasible hydropower potential has been exploited in Africa, 18 percent in East Asia and the Pacific, 21 percent in the Middle East and North Africa, 22 percent in Europe and Central Asia, 25 percent in South Asia, and 38 percent in Latin America and the Caribbean. Projects representing approximately 7 percent of total potential are currently under construction, of which the World Bank Group's share is below 10 percent (World Bank 2009).

IEG identified 211 projects that involved dams, whether for hydropower or other uses, in the evaluation portfolio. Of these, 100 projects involved hydropower, 57 of which were multipurpose projects. Almost a third (66) of the 211 projects were primarily rehabilitation projects. Many dams face gradual deterioration from lack of maintenance, and a number have been shut down because of salinity, sedimentation, and other problems.

The performance of the Bank's dams and hydropower projects as a group is on a par with that of other water subsectors.

The overall performance of the Bank's dam and hydropower portfolio is on par with that of the entire water portfolio: 77 percent of the 103 closed projects achieved an outcome rating of moderately satisfactory or better. (See appendix H for details on ratings.)[8]

BOX 5.4

HYDROPOWER AND MULTIPURPOSE PROJECTS

The Bank's 2003 water strategy notes that although the Bank has been important in attracting critical private sector investment in projects, there is growing concern that, "by disengaging from difficult, complex issues [that is, dams], the Bank is losing its credibility as a full-service investment and knowledge partner" (World Bank 2003b, p. 3). In response to this criticism, the Bank has been reengaging in hydropower.

The new hydropower development business plan includes rehabilitation of existing plants and the construction of small and run-of-river plants (plants that do not create a reservoir) and multipurpose plants with reservoirs. It supports feasibility studies for technically, economically, and environmentally satisfactory projects. The tendency is to move toward multipurpose projects that produce hydropower but also support irrigation, flood protection, or industrial use. In fiscal 2007 nine hydropower projects (five of which are "carbon-financed," that is, paid for in part by the sale of carbon emissions reductions; see IMF and World Bank 2007) were approved for a total of $748 million in new World Bank Group financing, including $115 million in guarantees and $66 million in carbon finance.

The Bank's water sector specialists see this "new" hydropower as demanding more sophisticated integration across disciplines, across water uses, across broader energy and water resources opportunities, across stakeholders (local and international), and across lending, reform, and capacity building. The new emphasis involves more projects that address both water supply and energy security.

Source: IMF and World Bank (2007).

Summary

Irrigated land accounts for about 80 percent of all water use. A recent IEG evaluation found that although the use of water for irrigation had increased dramatically in the developing world, attention to agricultural water resources management had decreased steadily, as had attention to environmental issues in projects dealing with agricultural water use. The discussion of agricultural water charges almost disappeared from Country Assistance Strategies.

Water user associations have been promoted as a vehicle for maintaining and operating irrigation and drainage schemes, but establishing them has been challenging: they were established as planned less than half the time, and only three-quarters of WUAs that were established appeared to work effectively at project closure. In the remainder, members received insufficient training, and the creation of the WUA generated little sense of ownership. Some WUAs failed to achieve long-term sustainability because strong technical or financial capacity was lacking.

Water services in urban areas are often subsumed under other rubrics and consist of more than just water supply and sanitation: only about half of projects that dealt with urban water did so exclusively. The success rate of urban projects has steadily improved, reaching 80 percent satisfactory in 2007. Nearly 30 percent of completed urban water projects dealt with water risk either directly or indirectly. Projects that explicitly address risk are generally concerned with preparing for or recovering from natural disasters, as well as addressing water quality issues or future scarcity. These projects have performed somewhat below the rest of the urban water portfolio.

The evaluation portfolio includes 218 projects dealing with rural water supply. These projects have generally increased access to safe water and provided a variety of benefits to rural families. Progress on access in China and India alone will be sufficient to achieve the MDG target for water supply. However, additional efforts are still needed to ensure access to improved water sources in rural areas.

Progress on sanitation has been limited: 2.6 billion people—40 percent of the world's population—remain without access to improved sanitation. Effective collection of urban wastewater has become a critical problem in most developing countries. The Bank's support in this area has focused more on wastewater treatment than on ensuring household connections to public sewerage systems. About two-thirds of sanitation projects achieved successful results. When project achievements fell short of expectations, institutional weaknesses in the sanitation sector were frequently the cause. To facilitate low-income households' access to sanitation, the Bank has supported subsidies, particularly in rural areas. Despite the subsidies, however, poor households still struggled at times to meet the counterpart requirements.

After a peak in the mid-1990s, worldwide dam construction slowed. The Bank has recently increased its financing of dam construction, in many cases for multipurpose dams that provide hydropower. A new hydropower development business plan, "Directions in Hydropower," completed in 2009, covers rehabilitation of existing plants, small and run-of-river plants, and multipurpose hydropower plants with reservoirs (World Bank 2009). It also advocates the use of feasibility studies so that projects will be technically, economically, and environmentally appropriate. The multipurpose projects also often support irrigation, flood protection, or industrial use. The overall performance of the Bank's dam and hydropower loans is on a par with that of the water portfolio as a whole. Although the Africa Region has the highest number of hydropower projects of all Bank Regions, the continent's potential remains largely untapped.

Chapter 6

EVALUATION HIGHLIGHTS

- Public providers remain the norm in WSS, but where the private sector has gotten successfully involved in urban WSS, it generally expanded infrastructure and increased fee collection.

- A common theme in projects where private sector participation encountered difficulties was lack of a conducive regulatory and institutional framework.

- Common positive outcomes in a decentralized setting were successful installation or rehabilitation of infrastructure and improved service delivery.

- IWRM projects focused most often on institutional development but frequently promoted top-down changes.

- The Bank facilitated cooperation among riparians in 31 international basins and funded some groundbreaking work on transboundary groundwater issues.

Photo courtesy of Curt Carnemark/World Bank.

Water Management Institutions

The Institutional Aspects of Water

Most countries have committed themselves to increasing access to water and sanitation in order to achieve the MDG targets. But a variety of paths have been taken to achieve this end. Two typical tactics to improve service delivery and increase efficiency are privatization of water service provision and decentralization of services to lower levels of government.

This chapter documents the achievements and problems encountered with both tactics.[1] Another tactic uses a project-by-project IWRM approach to improve both water service sustainability and water provision. This chapter also discusses the Bank's current approach to building the institutions that promote cooperation around transboundary waters.

The Context for Private Sector Participation in WSS

In 1985 the World Bank began to shift its development focus from the public to the private sector. Private sector participation (PSP) was fostered at first to close the large gap in infrastructure financing for developing countries. An early and persistent problem was that attracting private investments in piped systems, treatment plants, and new connections to water and sewerage systems is difficult when the rules are not clear to all involved stakeholders. Country institutional and regulatory systems had to be made more transparent, although it also helped if incentives could be put in place to encourage efficient utility operation and cost-effective labor and maintenance practices. Thus, in 1993 the Bank strategy paper on water resources management detailed how the involvement of the private sector would be encouraged (World Bank 1993).

In 1997 a Board-endorsed action program, "Facilitating Private Involvement in Infrastructure," gave more attention to the financing mechanisms deemed most effective in leveraging private funding for infrastructure. Such activities were to take place in conjunction with the Multilateral Investment Guarantee Agency (MIGA) or the International Finance Corporation (IFC) (World Bank 1997; IEG 2002).

With the idea of making a lasting impact on water utility reform in developing countries, the Bank, in conjunction with IFC, developed various forms of engagement with the private sector, including concessions, leases, and management contracts (appendix box I.1).

Private Sector Participation in Urban WSS

Thirteen percent of projects dealing with urban WSS aimed to introduce PSP in the sector. Seventy of these projects have closed, and 46 had managed to involve the private sector as planned by the time of closure. Among the 24 projects that did not, Turkey and the República Bolivariana de Venezuela terminated the contracts prematurely;[2] 6 countries (Algeria, Argentina, Nigeria, Rwanda, Tunisia, and Uganda) cancelled only one or two out of several contracts; Bolivia cancelled all contracts; and in 6 other countries (Guinea, Jordan, Kosovo, Sierra Leone, Trinidad and Tobago, and West Bank and Gaza), private companies efficiently managed service delivery for several years, but when the contracts came up for renewal, the government was reluctant or contractors were not ready to continue work. In some instances the areas where the work was to have been done had become conflict-ridden. In all these cases, water services management reverted to the public utility. In countries without the financial means to pay the operator, contract expiration meant that the water system continued to deteriorate, as was the case in the Trinidad and Tobago Water Sector Institutional Strengthening Project (P037006), for example. In other countries consumers hardly noticed the difference once the private operator left, as was the case with the Turkey Antalya Water Supply and Sanitation Project (P009093). The public utility proved able to continue service provision at the same level as the private operator.

Two-thirds of projects that aimed to facilitate PSP managed to involve the private sector as planned.

Several factors contributed to the difficulties of private sector involvement. A single project may have encountered several of these:

- Lack of effective regulation (27 projects)
- Civil unrest, conflict, or coup (14 projects)

- Natural disasters, such as floods, droughts, fires, and earthquakes (9 projects)

- Financial crisis (7 projects)

- Loss of private operator interest[3] (5 projects)

- Change from a government that promoted PSP to one that did not (4 projects).

Lease/*affermage* contracts (see definition in appendix box I.1) seemed the least affected by these factors, followed by management contracts and concessions. A well-functioning and well-maintained regulatory system (one able to control facilities and set or change prices) was important in determining whether a private operator was able to take over the provision of water supply or sanitation, although the creation of a new regulatory agency with little institutional capacity but a lot of discretionary power has been a major cause of the failure of PSP in several countries.

A supportive regulatory system has proved important to private operation.

World Bank self-evaluation reports and IEG assessments for the completed projects found that when a project successfully involved the private sector in urban WSS, it generally resulted in increasing both infrastructure provision and fee collection. The different contract types yielded slightly different results (see appendix tables I.1 to I.4). Concession contracts and management contracts resulted in greater efficiency gains for fee collection, tariff increases, and more installed water meters, whereas lease contracts focused slightly more on increasing the supply, production, and quality of water.

The private sector was generally successful in increasing infrastructure and fee collection in urban WSS.

Private Sector Participation in Rural WSS

Of 218 rural WSS projects, 56 aimed to encourage PSP. Those projects (34 completed, 22 ongoing) were spread over 41 countries, 12 of which had more than one operation (see appendix figure I.5). The Bank's role in this process was to conduct high-level discussions to promote PSP. The Bank also organized conferences to train government officials

in the preparation of contracts and bidding documents for PSP and funded studies to explore opportunities for PSP in the country context (see appendix table I.5). Bank support for the private sector has a notably long record in Africa and Latin America (also see appendix figure I.5).

The private sector's role is much more limited in rural areas.

Of the 34 completed rural projects that sought to introduce PSP, 29 succeeded in effectively getting the private sector involved in rural WSS; in 2 of the 5 projects that did not introduce PSP, legislative and institutional reforms set the stage for private participation, but PSP did not get off the ground during the life of the project. Among the completed rural WSS projects with PSP, projects in Albania, Costa Rica, Ethiopia, Madagascar, and Paraguay set up a regulatory agency or otherwise improved the institutional framework. However, a regulatory agency did not always prove necessary. Especially in countries with weak institutions, it was sometimes more effective simply to formulate contracts with the private sector and enforce their terms rather than change the country's whole regulatory system (Kauffmann and Perard 2007, p. 15). Thus, in Benin, Guyana, Kosovo, and the República Bolivariana de Venezuela, contracts were developed without a regulating agency in place. In 17 countries legislation was passed and policies were developed to allow for PSP in the water sector. Among those countries, Albania, Ethiopia, Ghana, and Rwanda also developed a strategy or action plan, mapping out a route to involving the private sector in the provision of rural WSS. In those countries the projects closed before PSP could be implemented.

International private firms involved in urban areas in developing countries were engaged in rural areas in only four countries (Burkina Faso, Kosovo, Rwanda, and Trinidad and Tobago). In all other projects, the local private sector was involved. The most common way to involve the private sector in rural water delivery was through the construction of rural water supply systems, followed by operation or maintenance activities (see appendix table I.6). For example, in the 1998 Madagascar Rural Water Supply and Sanitation Pilot Project (P001564), private operators signed 18 medium-term *affermage* contracts with the local government for the operation and maintenance of water systems. Over the course of the project, these operators expanded the

Photo courtesy of Dominic Sansoni/World Bank.

large amounts of funding in infrastructure investments or in increasing the efficiency and cost-effectiveness of water utilities. In rural areas the local private sector got involved in the operation of water systems, but it invested less and shared little financial risk. Its role was therefore much more limited and focused on efficiencies in management. What projects involving the private sector in rural WSS can bring in addition to water delivery is employment and an upgrading of local skills, both of which can help reduce rural poverty. Thirty-eight percent of projects actually created employment.

The international private sector tends to focus on large and medium-size cities; small-scale local operators can have comparative advantage in more remote areas, but their capacity is often limited.

Overall, although the Bank's attempts to increase private involvement in water service delivery had some success (see appendix I), the formal private sector has not played the investment role that had been hoped for, and public providers remain the norm (Marin 2009). Informal, small-scale private providers continue to play a large role in water supply to the poor in the cities of many developing countries, a role that the Water and Sanitation Program (a multidonor program administered by the Bank) and the Bank itself have played a constructive role in documenting.

Private involvement has been less successful in irrigation and drainage. According to a 2007 study undertaken by the Bank's Water Sector Board, "experience with public-private partnership in I&D is scant. Regarding client benefits in the public-private partnerships studied, the general result is improved but more expensive water service because of decreased government subsidies not fully compensated by efficiency gains" (World Bank 2007a).

Public providers remain the norm for service delivery.

service (mostly by supplying pipes, equipment, and meters) to 115,000 people in 24 small towns. A beneficiary survey of the project found that all 200 communities were managing their water systems satisfactorily with the support of the village caretakers trained by the project.

Engaging the private sector in rural areas often foundered on issues related to economies of scale (appendix box I.2). Water and sewerage companies in large and medium-size cities are attractive targets for the private sector, but international firms tend to ignore smaller, more remote towns and places where political interference is more prevalent (appendix boxes I.3 and I.4). Indigenous, small-scale private operators can have comparative advantage in such settings. The Bank can be useful in developing a favorable environment for such cases, as happened in the Colombia Water Supply and Sewerage Sector Project (P006836, appendix box I.3).

Comparison of rural and urban PSP (appendix I) reveals that where Bank support succeeded in helping countries bring the private sector into WSS, the international private firms tended to become involved in urban areas and contributed

Decentralization

Decentralization transfers authority and responsibility for governance and public service delivery from a higher to a lower level of government (IEG 2008c).[4] Of the 206 projects that either dealt with decentralization of services or operated in a decentralized setting (105 completed, 101 ongoing), 27 had decentralization of water service provision as an objective (19 completed, 8 ongoing). Of the remaining projects, 114 involved at least one decentralization-related component, and 65 were only carried out in a decentralization setting without a particular component directed at the decentralization process.

Most projects done in a decentralized setting involved capacity building, particularly at the local level.

These projects supported a combination of physical investments and policy and institutional changes. Most water supply and sanitation projects done in a decentralized setting sought to build local or central capacity, with local involvement clearly more prevalent. In terms of physical infrastructure, the largest number of projects focused on new water supply infrastructure, followed by construction of sanitation or sewerage infrastructure. Over 50 percent of projects had a cost recovery component (appendix figures J.29 and J.30).[5] Seventeen percent of projects attempted to strengthen local resource mobilization by adjusting the tariff system, the objective being to finance at least operation and maintenance costs through user contributions.[6] Thirty percent of projects also worked on improving operation and maintenance of WSS infrastructure, and another 30 percent developed an institutional framework for decentralization. Less frequently, but still in over 20 percent of cases, projects addressed environmental issues such as pollution, water

quality management, and water resources management, including IWRM. Finally, 20 percent also helped establish local water utilities in urban areas and WUAs in rural areas.

Under the right conditions, projects nearly always achieved some positive results, but they rarely met expectations fully.

Completed projects in a decentralized environment rarely met expectations fully, but when the budget and authority assigned to the lower level of government matched the responsibility delegated to it, the projects almost always had some positive achievement. Projects were most successful with the construction or rehabilitation of infrastructure and improvement in service delivery. In general, activities such as maintenance and cost recovery, which are a challenge outside of a decentralized context, were challenging within it as well. Only about one-third of projects that aimed at improving operations and maintenance and one-third of projects aiming at improving cost recovery succeeded in doing so. About half of projects that aimed to strengthen local capacity and two-fifths of projects that supported institutional reforms were successful. Other positive outcomes usually associated with decen-

FIGURE 6.1 Activities of Completed Decentralization Projects

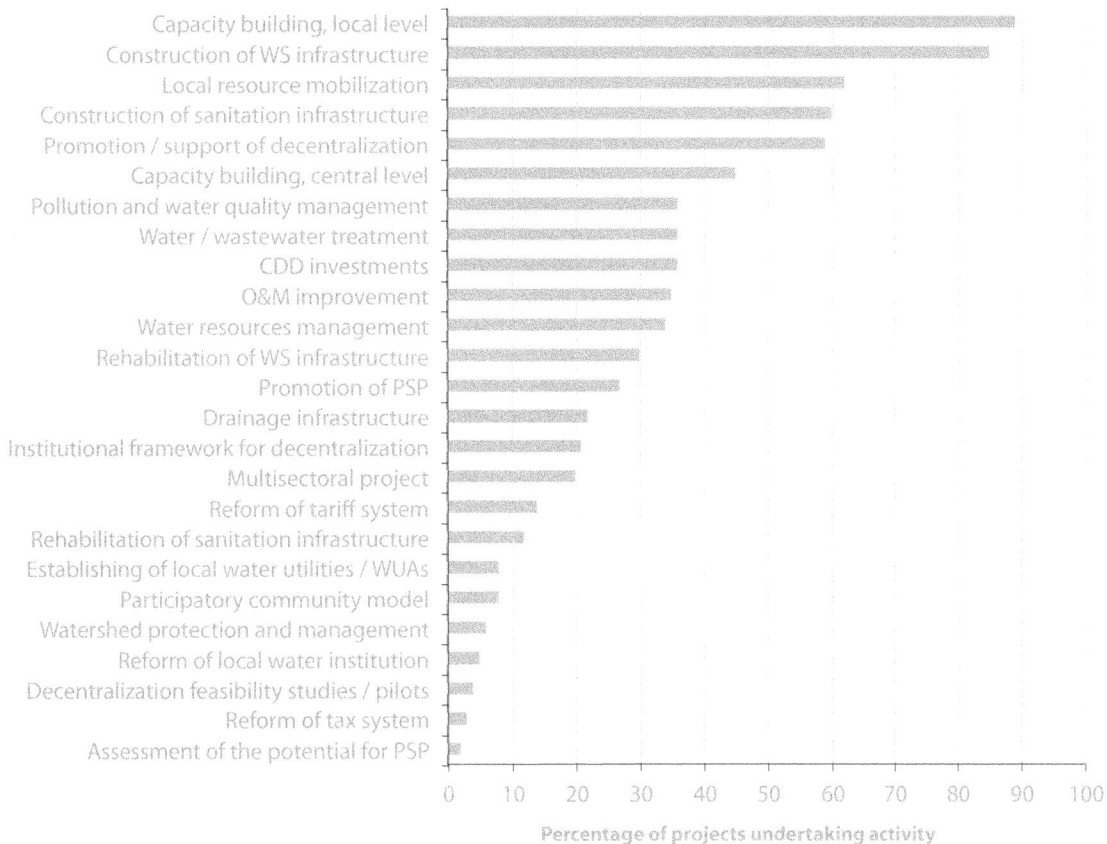

Percentage of projects undertaking activity

Source: IEG water database.

Note: CDD = community-driven development.

tralization—increases in accountability, ownership, empowerment, and social cohesion—were achieved in a minority of cases, although for about half of the relevant projects at least one of those elements was reported to have improved (figure 6.1). An incomplete decentralization process was associated with negative project outcomes in over 10 percent of cases, suggesting that it is important that decentralization not lead to unclear or overlapping responsibilities and, in particular, that administrative decentralization be followed by transfer of the respective resources. Box 6.1 summarizes the experience with several water projects in a decentralized context.

Figure 6.2 depicts the success rates of the decentralization process by decentralization type for the aggregate subset of 105 completed projects.[7] Decentralization occurred as planned in 61 percent of the closed projects; in 12 percent it did not, and decentralization was partial in the rest. As can be seen from this disaggregation, projects carried out in a devolution framework had the highest success rate (where success is defined as in table 6.1).

Decentralization was successful in 61 percent of projects—but some types of decentralization were more likely to succeed than others.

Regression analysis of the factors that influence decentralization success (see appendix table J.39) found that in addition to decentralization type, important explanatory factors include the adequacy of cost recovery and whether decentralization was neglected or incomplete (for example, with no particular component directed at supporting the process even though implementing capacity is still weak, or with no fiscal decentralization). This implies that capacity building, in particular with respect to financial management (since that was a frequent focus of these efforts), is essential, especially if local capacity is still weak and if decentralization is less extensive, such as in deconcentration settings. The findings also suggest that neglecting the decentralization process can lead to undesirable negative outcomes, which implies that, in a decentralization setting, how the decentralization process can be supported and how the institutions involved can be strengthened should always be evaluated.

Delegation projects seem to have been only slightly less successful than projects aimed at devolution, but projects carried out in a deconcentration setting performed worst. Hence, the type of decentralization seems to affect the success of a project.[8] Since devolution is the deepest form of decentralization, the finding also suggests that it should be easier to elicit the potential benefits of decentralization if authority is transferred completely, provided there is adequate local capacity. Less extensive forms imply potential coordination problems when critical aspects of the decentralization process remain incomplete, leading to an imbalance between responsibilities and authority or resources.[9]

BOX 6.1

EXAMPLES OF DECENTRALIZATION PROJECTS IN THE WATER SECTOR

The Morocco Rural Water Supply and Sanitation Project ((P040566, completed in 2003) sought to shift from the conventional supply-driven approach to service provision to a demand-responsive, community-based approach that involved setting up village-level WUAs to participate in the design and to take over O&M of the water supply schemes. The project included the creation, training, and operation of social mobilization teams (SMTs) responsible for the implementation of the participatory approach and hygiene education. However, training by SMTs was limited to social mobilization. Training on O&M, financial management, and hygiene generally was not provided. In practice, SMTs concentrated on getting construction under way and provided little support to WUAs once the schemes were operational. As a result, each scheme is managed by a WUA, but where social capital is weak, these associations are fragile and need prolonged support from SMTs until they can manage the systems on their own. In particular, they did not receive help to set up their tariff structure. Hence, the price of water usually covers only operational expenditures and not replacement costs, as was expected at appraisal.

The Lao PDR Luang Namtha Provincial Development Project (also completed in 2003) promoted decentralization of management responsibilities at the provincial level. However, the insufficient transfer of resources from the central government and limited mobilization of local revenue resulted in inefficient allocation of resources. Heavily dependent on the central government for resources, the provinces face constraints in developing any project on their own. Moreover, existing legislation does not provide incentives to the provinces for collecting or generating new sources of revenue for investments. This could undermine future efforts to strengthen local capacity and ensure the overall effectiveness of decentralization. Concrete steps need to be taken to increase the transparency of the funds being transferred to the provinces from the central government, improve the reliability of transfers in terms of timing and amount, and enhance planning and decision-making capacity at the provincial level.

Sources: Project ICRs and IEG water database.

FIGURE 6.2 Success of Decentralization by Decentralization Type

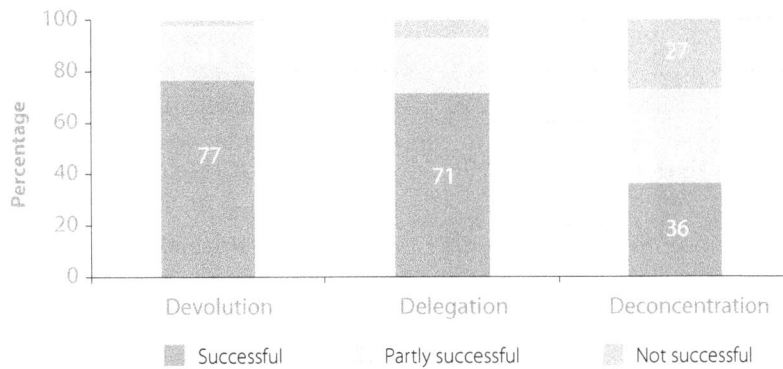

Source: IEG water database.

Note: Only those projects where decentralization affected the outcome are included. The total number of projects was 90 (43 devolution, 14 delegation, 33 deconcentration). For statistical significance levels see appendix table J.39.

TABLE 6.1 Classification of the Success of Decentralization

Decentralization was successful if...	Decentralization was partly successful if...	Decentralization was not successful if...
Service delivery improved while none of the other indicators outlined in box 6.2 deteriorated.		
At least two of the other indicators improved while service delivery as well as the remaining indicators stayed constant.	One of the other indicators improved while service delivery and the remaining indicators stayed constant.	
Service delivery improved or stayed constant while at most one of the other indicators deteriorated.	In any other case.	

Source: IEG data.

Delegation projects had an only slightly lower success rate than devolution projects; projects in a deconcentration setting were least successful.

Integrated Water Resources Management

Following internationally accepted best practice, the Bank has endorsed the IWRM approach (see box 1.1 in chapter 1) in the water sector, beginning with its 1993 Water Resources Management Policy Paper. Implementation of this approach was facilitated by the Bank's shift to matrix management in the late 1990s. This integration was further reinforced by the reorganization of the water sector in 2007, which created a single Water Sector Board (see appendix

Photo courtesy of Simone D. McCourtie/World Bank

C). The Bank continued its support for the IWRM approach in its 2003 strategy.[10]

IEG identified a total of 125 projects whose design and implementation were wholly or largely consistent with IWRM principles. These projects focused most frequently on institutional development activities, including financial and regulatory capacity building. Some countries (India, Kosovo, Mexico, Morocco, the Philippines, and Tanzania, among others) passed water laws that embodied IWRM principles. In addition, river basin management institutions were often set up to develop basin plans and make water allocation decisions involving relevant stakeholders (box 6.3). In some cases there were attempts to administer and enforce water rights, including through the collection of water fees. Not all the IWRM-consistent projects focused exclusively on institutional reform, however. Others combined activities such as setting up hydrometeorological monitoring systems with the provision of irrigation or water supply infrastructure (see also chapter 3 on river basin organizations).

The 1998 China Second Tarim Basin Project (P046563) illustrates how IWRM principles were implemented in the context of irrigation improvements. The project effectively brought about institutional reforms at the river basin management level. It also introduced a water quota system by establishing restrictions on agricultural water use, increased the productivity of water usage in irrigated agriculture, and established a water quota allocation specifically for a downstream riverside environment, which resulted in significant environmental water releases.

Discussions with water professionals from both borrower countries and donors, together with the evidence acquired during the case study missions to Brazil, Morocco, Tanzania, Vietnam, and the Aral Sea, indicate that much of what has been accomplished with IWRM in the water sector can be done in a top-down manner without greatly changing what is happening on the ground, such as the writing of laws and regulations, but significant provisions for enforcement are generally not a priority. Although some new institutions have been created, more have simply been renamed (from *Irrigation* to *Water Management*, for example). To some degree the approach to implementing IWRM has been formulaic, and this has impeded success (see box 6.4 for an example from Morocco). The most obvious manifestation has been the promotion of river basin organizations without taking the political context sufficiently into account.

Countries often make desperately needed changes to the way they manage water resources in response to a shock, such as a natural disaster or other major catastrophe such as a spill of toxic pollutants into an important drinking water source for a major city. When Morocco confronted a severe drought during the 1990s, decision makers at the highest political levels decided that efforts to reduce demand should supplement the search for increased water supplies.

Similarly, in Tanzania the 2004–06 drought, which saw economic growth decline by over 1 percentage point, was deci-

BOX 6.3

DATA ON WATER AVAILABILITY HELP BUILD CONSENSUS

The 2001 Kosovo Pilot Water Supply Project (P070365) carried out a water availability study that became the basis for water allocations within the watershed. The primary competing users in the project area were domestic water users (households) and irrigators. The study provided a rational basis for water allocations based on technical criteria such as the agronomy of the irrigation needs, the demands for domestic water, and the environmental flow requirements.

Source: Project ICR.

sive in moving the country further along the path to IWRM already established by a 1981 law (World Bank 2006c).

Evaluation missions also saw major progress made with IWRM after shock events in pollution-menaced Brazil, and in Vietnam after a major release of industrial chemicals into a critical river basin. A fast-onset crisis that compromises the water supply of a major city, such as in Brazil, when pollution threatened Brasilia or the drought put the survival of Fortaleza at risk, is more likely to provoke dramatic action.

But if the problem moves slowly, as in the Republic of Yemen, where the water table drops year after year, change, too, happens very slowly.

Transboundary Water: Managing across National Frontiers

As an international financial institution, the World Bank has a long tradition of fostering transboundary cooperation around projects affecting international waterways.[11] The Bank can assist its borrowers in addressing transboundary issues only where there is a willingness to negotiate. A stark divergence of national interests, a weak regional institutional and regulatory framework for water, or political instability can represent a major constraint and prevent the resolution of disputes within a transboundary river basin.

Cooperation on International Waterways

The evaluation identified 123 Bank-funded projects with activities related to international waterways, 40 of which were financed by GEF grants. During the period reviewed, 27 projects were closed, and the remaining 96 are still ongoing, pointing to the increasing importance of transboundary projects in recent years. The amounts committed to transboundary projects were $6.2 billion in mostly IDA and IBRD funding and $273.5 million in GEF grant funding. Almost 70 percent of the lending for transboundary projects went to the Africa Region and the Europe and Central Asia Region (50 and 45 projects, respectively; figure 6.3), the two Regions with the most transboundary basins.

To discern whether the Bank—as stipulated by its strategy—is focusing on those watercourses where its support is most likely to have an impact, the evaluation team ranked river basins by the number of countries sharing the basin and compared that ranking with a ranking of basins by number of projects undertaken in the basin. The analysis suggests that the Bank does prioritize those international waterways that are shared by the largest number of countries (appendix table J.40). Overall, the number of riparian countries per project ranged from 2 to 22, with a median of 4. In total, the Bank facilitated cooperation among riparian countries in 31 international basins, but not all riparians in all basins participated in project-funded work.

World Bank– and GEF-funded projects fall into three categories: large-basin initiatives; support for the implementation of treaties and conventions, including projects that foster cooperation around topics such as pollution, biodiversity, and groundwater conservation; and projects with activities largely determined by the provisions of the Bank's operational policies—that is, those that triggered a Bank safeguard.

The Bank has operated as a go-between and taken a flexible approach to supporting cooperation among riparians sharing a basin.

Large-Basin Initiatives

The Bank has supported cooperation among riparians, often in multiproject programs, focusing on tasks centered on

---

FIGURE 6.3 Transboundary Projects and Their Total Lending by Region

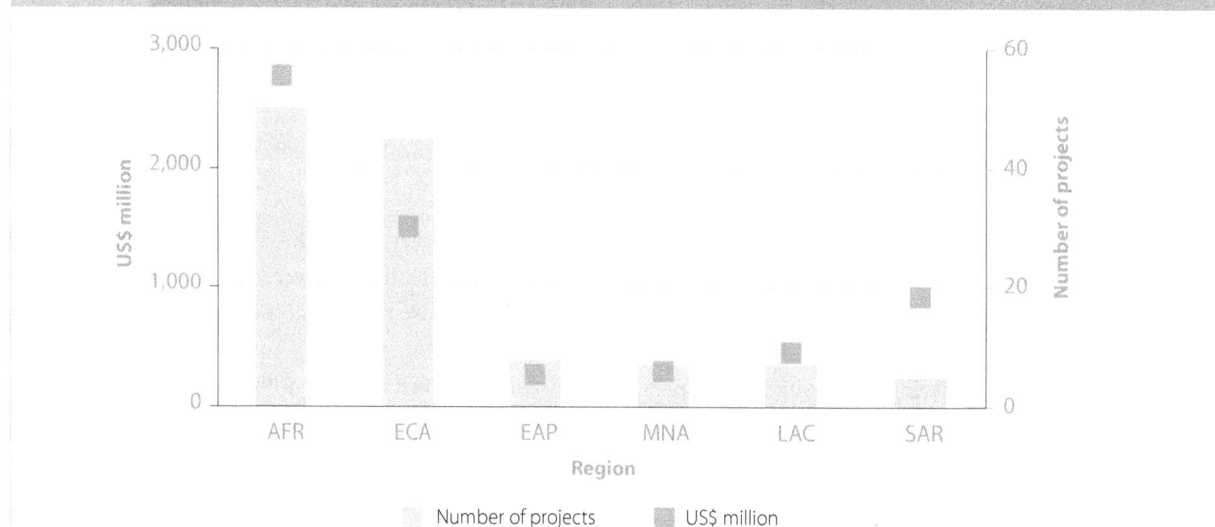

Source: IEG water database.

Note: AFR = Africa; ECA = Europe and Central Asia; EAP = East Asia and Pacific; MNA = Middle East and North Africa; LAC = Latin America and the Caribbean; SAR = South Asia.

THE NILE BASIN INITIATIVE

During the study period, a total of 19 projects involved the Nile Basin, including Lake Victoria. In 8 of these projects the Bank facilitated cooperation for environmental restoration or power deals between countries; in the other 11 the Bank notified other riparians of projects in the basin.

The Bank has been supporting the Nile Basin Initiative (NBI) since 1997, when the Nile Council of Ministers (Nile-COM) first requested support to coordinate donor involvement and establish a consultative group to raise financing for cooperative projects. The NBI was formally launched in February 1999 by the ministers of water affairs of the 10 countries that share the Nile Basin: Burundi, Democratic Republic of Congo, the Arab Republic of Egypt, Ethiopia, Eritrea, Kenya, Rwanda, Sudan, Tanzania, and Uganda. Together, these ministers make up the Nile-COM. The NBI is guided by the countries' shared vision "to achieve sustainable socio-economic development through the equitable utilization of, and benefit from, the common Nile basin water resources."

The basin-wide program is intended to build capacity and confidence. It also involves subbasin programs of investments to demonstrate the development results of cooperation. Despite regular meetings between riparians, tensions between them persist, such as between Uganda and Tanzania around water levels in Lake Victoria, which have been falling dramatically in recent years. Following two dry years (2004–05) when little hydropower could be produced, the Bank helped Uganda, through the 2006 Uganda Thermal Power Generation Project (P069208) , to find alternative means to generate electricity in order to reduce friction between the riparians. In parallel, and although progress is slow, the Bank is helping countries to align their priorities with the regional perspective through the NBI.

Source: "The World Bank and NBI," go.worldbank.org/C25RHXSYG0, retrieved on November 11, 2009.

large basins. An early example is the Indus Water Treaty of 1960 between India and Pakistan, where Bank involvement led to successful settlement of a major international disagreement over a river basin. During the 10 years of negotiation, the Bank's role as an impartial go-between made agreement possible (World Bank 1998, p. 155). During the period studied for this evaluation, the Bank has facilitated cooperation agreements along international waterways in the Congo, Lake Chad, Niger, Nile (box 6.5), Senegal, Volta, and Zambezi Basins in Africa, as well as in the Aral Sea, Caspian Sea, Danube, La Plata, and Mekong Basins, among others. Results have been mixed. Although the organizations created are generally fragile and often confront political issues beyond their capacity to resolve, the mere fact that an institution now exists to deal with the issues that arise is clear progress.

Treaties, Conventions, and Institutional Strengthening

The Bank's 1993 strategy paper recommended technical, financial, and legal support to help governments establish or strengthen river basin organizations addressing transnational water management activities. In 35 projects the Bank facilitated international agreements, charters, protocols, and memoranda of understanding addressing cooperation around international waters. Examples are the Senegal River Water Charter, where Guinea was included as a fourth riparian, and the Lake Victoria Protocol. In addition to these agreements, international river basin institutions were organized for the Aral Sea, the Nile Basin, the Senegal Basin, Lake Victoria, Lake Chad, and the Lielupe River shared by

Lithuania and Latvia (see also the discussion on river basin management in chapter 3). These institutions create a forum where representatives from different countries meet and discuss water allocations and water quality issues. Creating an enabling environment that fosters cooperation and openness to lending opportunities is as important as the lending itself.

The Bank facilitated agreements, charters, protocols, and memoranda of understanding in 35 international waters projects.

Photo courtesy of Arne Hoel/World Bank.

During the evaluation period, a water treaty was signed between Lesotho and South Africa for a scheme that transfers water to the latter. In addition to brokering the deal, the Bank provided funding for the transfer infrastructure. Another agreement was reached between Azerbaijan and the autonomous Russian region of Dagestan. Under the agreement, infrastructure in Azerbaijan regulated water flows so that 20 percent of the annual flow of the Samur River was diverted to Dagestan, 61 percent to Azerbaijan, and 19 percent kept in the river for environmental and other needs.

The subset of transboundary projects in the evaluation database also addressed transboundary data collection and monitoring activities as well as institutional capacity building (box 6.6). A review of project experience indicates that institutional strengthening was the most difficult activity in the transboundary projects.

Institutional strengthening has proved to be the most difficult activity in transboundary projects.

Twenty-seven projects addressed pollution from oil spills and ship waste (box 6.6). The outcomes of 11 closed projects were rated satisfactory. In these projects the Bank also facilitated the signatory process and, once it was completed, helped countries to comply with the relevant international conventions, for example, the International Convention for the Prevention of Pollution from Ships (MARPOL 73/78). One example was the Wider Caribbean Initiative Project (P006956, closed in fiscal 1998), which provided support in the ratification and implementation of MARPOL to 22 Caribbean countries. Another example was China's Ship Waste Disposal Project (P003630), whereby, as a result of waste treatment systems installed at 6 ports, over the next 20 years an estimated dumping of 6.3 million tons of waste

per year will be averted. This was a major achievement for the relatively modest sum of $17 million provided by the GEF-funded project.

In 27 projects that addressed oil spills and ship waste, the Bank facilitated the signatory process and helped countries comply with international conventions.

Transboundary Groundwater Management

World Bank projects for transboundary aquifer projects are rather limited in number, but the Bank has funded some groundbreaking work. The Bank's flagship project in transboundary groundwater was the 2002 Guarani Aquifer Project (P068121), funded by the GEF in the amount of $32 million, which involved all four riparian countries: Argentina, Brazil, Paraguay, and Uruguay. The project funded a transboundary assessment of the age and quality of the Guarani aquifer. It also created institutions in each country, organized a public information campaign, and supported a few cross-border subprojects. According to a project-level evaluation report, as a result of the project, the risk of over-extraction and contamination of groundwater resources and any transboundary impacts have been significantly reduced.

The Bank has funded some groundbreaking work on transboundary groundwater.

Inland Water Transport

Navigation does not figure largely in the current transboundary portfolio, but Operational Policy and Bank Procedure (OP/BP) 7.50 lists navigation as one activity requiring notification.

BOX 6.6

THE GULF OF AQABA ENVIRONMENTAL ACTION PROJECT

The 1996 Gulf of Aqaba Environmental Action Project (P005237) in Jordan improved environmental management of that body of water and fostered transboundary cooperation between Jordan and Israel. Project teams on all sides were determined not to allow politics to impinge upon environmental cooperation. As a result of the project, Jordan is now equipped with financial and productive human resources to implement public awareness programs, coastal policing and enforcement, marine pollution prevention and response, and marine park management. Cost recovery mechanisms were put in place and include marine park fees (diving fees, visitor fees, and beach facility fees); issuance of permits (air emission permits; cooling water discharge permits; resource user fees for import and export of all goods, to be set at 0.05 piaster per ton; and a 25 percent surcharge on the use of port reception facilities once they exist); and fines for environmental damages, including industrial pollution and oil spills. All revenue from these fees and fines is earmarked for the Department of Environment, Regulation, and Enforcement and contributes to project sustainability.

Source: Project files: Jordan—Gulf of Aqaba Environmental Action Project.

THE INLAND WATERWAYS AND PORTS PROJECT IN THE MEKONG RIVER DELTA

The Inland Waterways and Ports Project (IDA-30000; P004843) in Vietnam contributed to a huge increase in waterway traffic and transported cargo, although the extent to which the throughput is international is not known. In 1995, 23.5 million tons of cargo was carried through the inland waterway. By 2004, the last year for which data are available, cargo transported had risen to 55 million tons. The project financed major improvements to Can Tho Port, including paving the entire port area, installing dock bumpers and other improvements to berths, and adding cranes for container loading and unloading. The increased ship traffic in the improved canals and waterways is reflected in a dramatic increase in the port's throughput.

During project preparation, Can Tho's cargo flow was originally predicted to increase to about 1.2 million tons by 2010. By late 2008, however, total cargo flow was already at 3 million tons, about six times what was to be expected at this stage. And the profitability of Can Tho Port had grown sixfold.

Source: Project Performance Assessment Report.

Port and navigation infrastructure have been important in facilitating trade between countries (box 6.7). Fifteen projects in the water transport portfolio overlap with projects in the transboundary waterways portfolio. In interviews, some Bank staff mentioned that staff working on water issues and those working on inland water transport do not communicate regularly. In addition, the Bank has produced few publications on inland water transport. Supporting countries' investments in inland water transport and port facilities may be another approach to fostering their cooperation along international waterways.

Transboundary projects have not taken advantage of the linkages to ports, navigation, and trade.

Summary

Although the Bank's attempts to increase private involvement in water service delivery have had some success, public provision remains the norm. Where private provision of WSS services was successfully introduced, efficiency often improved. Comparison of rural and urban private sector participation indicates that where Bank support succeeded in helping countries bring the private sector into WSS, the international private sector tended to become involved in urban areas and contributed large amounts of funding for infrastructure investments or to increase the efficiency and cost-effectiveness of water utilities. In rural areas the local private sector got involved in the operation of water systems, but it invested less and shared little financial risk, focusing mainly on improving efficiencies in management.

In many countries, public water service delivery is being decentralized to increase the capacity for service delivery and improve efficiency. When the budget and authority allocated to the lower level of government matched the responsibility assigned to it, projects operating in a decentralized setting almost always generated some positive results, even if they seldom fully met expectations. Projects were most successful with the construction or rehabilitation of infrastructure and improvement in service delivery, but progress on improving operations and maintenance and strengthening cost recovery was limited. About half the projects that aimed to strengthen local capacity and two-fifths of those that supported institutional reforms were successful.

The Bank adopted the IWRM approach to the water sector beginning with the 1993 Water Resources Management Policy Paper. The 125 projects that were wholly or largely consistent with IWRM principles focused most frequently on institutional development. The spread of IWRM to the Bank's borrowers has been slow, and it appears that countries are most likely to make needed changes after a shock event, such as a major catastrophe. The evaluation has seen at least one water-stressed country where major progress has been made with IWRM after such an event, but where the water problem worsens slowly, change likewise happens very slowly.

The Bank has a long tradition of fostering transboundary cooperation around projects affecting international waterways, and it has funded an increasing number of transboundary projects in recent years, partly with GEF support. The Bank is a proven broker in transboundary issues and has appropriately given priority to international waterways that are shared by a large number of countries.

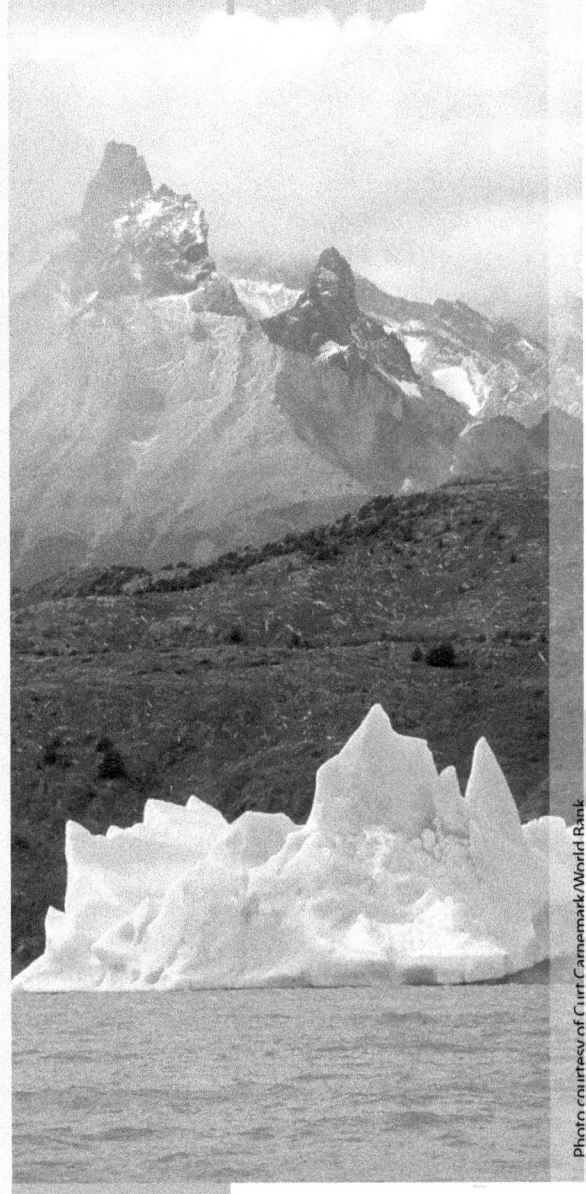

Conclusion, Findings, and Recommendations

Water has been a major focal area for World Bank lending from its earliest years. Through its support for investments in irrigation, water supply, sanitation, flood control, and hydropower, the Bank has contributed to the development of many countries and helped provide essential services to many communities. Bank staff have developed an extensive menu of options—this study identified more than 400 different water-related activities supported by Bank-financed projects—that can be used to put together support tailored to each borrower's needs and circumstances.

The 2003 Water Resources Sector Strategy developed 23 strategic objectives. This evaluation found evidence that achievements have been made under each of those objectives (appendix table J.42). Projects currently managed by the Water Sector Board represent less than half of the portfolio, reflecting the strategy's goal to integrate water into other sectors. IEG performance ratings track a steady improvement in sector performance. During the most recent five-year period, water showed the strongest performance improvement of any sector. Water sector ratings are now slightly better than the Bank-wide average, although they still trail those of several other sectors.

Projects have evolved in some of the right directions. For example, more attention is being given to wetlands, and the Bank has been more cautious about financing irrigation projects that rely on falling underground aquifers. Its approach to dams has become more balanced and more environmentally and socially sensitive, as is its commitment to "new" hydropower that takes these lessons into account. First steps with watershed management and water resources management are clearly moves in the right direction. Yet the Bank, its partners, and its clients have tended not to respond at scale to some increasingly pressing challenges, especially the preservation and restoration of groundwater resources, water quality improvement, coastal vulnerability, and sanitation.

Although the water portfolio has witnessed a steady improvement in performance measured against projects' stated objectives, the evaluation also found that the Bank and its clients have underemphasized some of the most challenging areas, such as fighting pollution, environmental restoration, and water quality monitoring. The limited success with full cost recovery has caused the Bank to moderate its approach, but the question of who will pay for uncovered costs remains unanswered. The evaluation shows that success rates in the more difficult areas have been lower, yet these areas will need to be addressed in the years ahead given that growing water stress will determine development outcomes in many countries.

Main Findings

Increased Lending and Improving Project Performance

The Bank increased its lending for water and the number of countries it served during the period evaluated. Although the number of countries that borrow for water has varied from year to year, 79 borrowers were served in 2007, compared with 47 in 1997. Lending for water increased by over 50 percent during the period.

The integration of water practice across Bank sectors appears to be well under way. Integration of water practice was an important goal of the 2003 Water Resources Sector Strategy, and during the period evaluated the majority of water-focused projects were overseen by sector boards other than Water Supply and Sanitation.

Water projects in the aggregate have good success rates when measured against objectives. IEG performance ratings show steady improvement in sector performance measured against project objectives. During the most recent five-year period, water was the most improved major sector by this criterion, with a particularly noteworthy 23-percentage-point improvement in the share of satisfactory projects in the Africa Region. Within the portfolio, 77 percent of the 857 completed projects had an aggregate outcome rating of satisfactory ("moderately satisfactory" or better), slightly above the Bank-wide average of 75 percent. The trend continued in 2008, when water sector projects attained a 90 percent satisfactory rate.

The focus of Bank activity has shifted over time. The Bank has lent heavily for irrigation and water supply, and dams and hydropower have become more important in the last few years. But some activities that are of growing importance as water stress increases have become less prominent in the Bank's portfolio—notably, coastal zone management; pollution control; and, to a lesser degree, groundwater conservation. Although the portfolio has performed well when measured against projects' stated objectives, the Bank and the borrowing countries as a group have not yet sufficiently tackled several tough but vital issues, among them sanitation, fighting pollution, restoring degraded aquatic environments, monitoring and data collection, and cost recovery. Where it has lent for hydrological and meteorological monitoring, the Bank has focused on providing technology for data collection and relatively less on gathering and interpreting information for which there is an identified demand. Such aggregate findings, however, mask Regional and country-specific variations and needs. For example, East Asia and Africa have responded more actively than other Regions to the sanitation challenge. These issues are covered in greater detail below.

Water Resources Management

Effective demand management is one of several critical challenges worldwide in the face of increasing water scarcity. Demand for water can be affected by three broad sets of measures: pricing, quotas, and measures to improve water use efficiency.

Efforts to improve the efficiency of water use and limit demand in agriculture, the largest consumer of water, have had limited success. Efficiency-enhancing technologies alone do not necessarily reduce on-farm water use, and efforts to manage demand with water charges in agriculture have encountered limited success, partly because of the low price elasticity of demand for agricultural water. Fixing and enforcing quotas for water use is a relatively recent approach and deserves careful evaluation when more projects featuring this approach have been completed. Cost recovery in Bank-supported projects has rarely been successful: only 15 percent of projects that attempted cost recovery achieved their goal. The successful projects have generally been those that improved the efficiency of water institutions at collecting fees. This limited success has caused the Bank to moderate its approach, but without clearly identifying sources to finance the recovery shortfall, threatening the sustainability of investments.

In the area of water supply, reducing UfW has been the main activity directed at improving water use efficiency. About half of projects that attempted to address UfW managed to reduce it by at least 1 percent. Finding effective ways to improve efficiency and manage demand for water will be critical if the Bank wants to maintain a leading role in this area.

Integrated water resource management, the focus of two consecutive Bank water strategies, has gained traction within the Bank but has made limited progress in most client countries. Within the Bank there has been considerable progress in integrating water into the work of other sectors and in consolidating institutional structures to carry out water-related activities. However, outside the Bank, even in countries where IWRM is now well integrated into the legal framework, it is known mainly just in the water sector. The information necessary to inform decision making is not easily available, and perhaps more important, the economic implications of water constraints are not widely appreciated. Meanwhile there are indications that the Bank is paying less attention to data collection, an essential prerequisite for successful IWRM implementation, because countries have less motivation to confront a situation with unknown parameters.

When IWRM is successful, it most often happens in a particular location at a time of necessity. Some countries have made progress with water resources management after natural disasters, for example. Shocks often do not affect entire countries, however, nor are they a desirable route to IWRM. The way to open the window of opportunity without waiting for a shock is to support monitoring processes that deliver information to relevant public and private stakeholders. The example of Brazil shows that making water data publicly available over the Internet helps increase stakeholder concern, which helps to mobilize the political will necessary to confront entrenched water problems.

The number of projects dealing with groundwater issues has been declining, but within that problematic trend the portfolio has also witnessed a positive shift away from a focus on extraction. This shift is important given falling water levels in critical aquifers in many Bank borrowers.

Within the groundwater portfolio, activities aiming to increase water supply were the most successful as a group, whereas activities related to reducing pressure on groundwater and promoting conservation generally proved more challenging. Yet such activities will need to become more prominent in the portfolio, if the Bank is to effectively help the growing number of water-stressed countries address increasing groundwater scarcity. In the Republic of Yemen, for example, rapidly growing consumption of irrigation water caused by improved tube well technology and the availability of highly subsidized diesel fuel has resulted in irrigation extracting over 150 percent of the country's renewable water resources.

Watershed management projects that take a livelihood-focused approach perform better than those that do not. Projects combining livelihood interventions with environmental restoration enjoyed high success rates, even though effects on downstream communities (such as reduced flooding and improved water availability) and social benefits in both upstream and downstream communities were often not measured. Hydrological monitoring (with or without remote sensing) and watershed modeling could help improve impact assessment, and thus make it easier to calculate the cost-benefit ratio of such interventions.

Environment and Water

Environmental restoration is underemphasized in Bank projects, possibly because its immediate and long-term financial importance is unclear. More attention to cost-benefit calculations could help the Bank and its clients evaluate trade-offs and get better results.

Most Bank projects focus on infrastructure, even though in some cases environmental restoration is more strategically important. It is not always necessary to restore the water-related environment to a pristine state in order to obtain major social, economic, and environmental benefits and reduce vulnerability. Priority improvements to degraded environments, even when small, can have big impacts. A coastal wetlands protection project in Vietnam, for example, successfully balanced reforestation with livelihood needs. The project successfully reforested critical areas and led to a substantial reduction in coastal zone erosion.

Countries and donors need to focus more on coastal management going forward, as some 75 percent of the world's population will soon be living near the coast, putting them at heightened risk from the consequences of climate change. Approvals of Bank projects in this area have dwindled over time, and the reasons for this should be considered in the Mid-Cycle Implementation Report.

Many projects contain funding for water quality management, but few countries measure water quality. The number of projects that actually measure water quality is declining. Evidence of improved water quality is rare, as are indications of the improved health of project beneficiaries. The data that are generated need better quality control. Water quality in the top five borrowing countries is declining, and fewer than half of projects that set out to monitor water quality could show whether an improvement had taken place.

Water Use and Service Delivery

The Bank has increasingly focused on water service delivery, but there has been a declining emphasis on monitoring economic returns, water quality, and health outcomes. Only a third of wastewater treatment and sanitation projects calculated economic benefits.

Sanitation needs greater attention. Population growth in developing countries has been rapid, as has urbanization. An expansion of piped water services and increased household water use will lead to an accelerating demand for adequate sanitation. The evaluation recognizes that even if the Millennium Development Goal for clean water supply is achieved, 800 million people will still lack access to safe drinking water in 2015, but a much larger number—1.8 billion—will still lack access to basic sanitation. Within sanitation, more emphasis is needed on household connections. Connection targets in projects are generally not met, and IEG has seen a number of treatment plants functioning below design capacity because households have not connected to the systems, in part because willingness to pay has been overestimated and facilities have been overdesigned. This report highlights the particular weakness of sanitation institutions, which will continue to constrain progress until their capacities improve.

Hydropower projects have performed well, and significant untapped potential remains for appropriate development, particularly in Africa. After a peak in the mid-1990s, dam construction around the world slowed. The Bank has recently increased its financing for dam construction, in many cases for multipurpose dams that provide hydropower and often also support irrigation, flood protection, or industrial use. Almost a third (66) of the 211 Bank-financed dam and hydropower projects covered in the evaluation rightly focused on dam rehabilitation, as many dams have experienced gradual deterioration due to lack of maintenance, and a number have been shut down because of salinity, sedimentation, and other problems. A new hydropower development business plan, "Directions in Hydropower," was completed in 2009 and supports feasibility studies so that projects will be technically, economically, and environmentally appropriate. Indeed, it will be vital to take on board the experience with hydropower projects, including their scale, socioeconomic, and environmental impacts.

Institutions and Water

Water services are delivered by public providers in most countries, but private sector participation has made some progress. Where international private firms have been successful in urban areas, they have contributed significant investments to infrastructure, and in some cities they have managed to increase the efficiency of water utilities' operations. In some Bank-financed projects in rural areas, the local private sector manages the operation of water systems, but it has invested little and shared little of the financial risk. Where governments want private involvement, a well-functioning, well-maintained regulatory system is necessary for its sustainable participation in utility operations. In many cases this has remained elusive, and this has limited private sector involvement.

Projects operating in a decentralized environment have had difficulty meeting expectations, but when the budget and authority accorded to the lower level of government have matched the responsibility assigned to it, projects have had positive achievements. Half of projects that aimed to strengthen local capacity and two-fifths of projects that supported institutional reforms were successful. Other positive outcomes usually associated with decentralization—increased accountability, ownership, empowerment, and social cohesion—were achieved in a minority of cases.

Support for institutional reform and capacity building has had limited success in the water sector. Institutional reform, institutional strengthening, and capacity building have been the most frequent activities in Bank water-related lending. Yet these interventions have often been less than fully effective, and weak institutions have often been responsible for project shortcomings.

The Bank has been actively engaged in addressing transboundary water issues. In its choice of transboundary projects, the Bank has given priority to international waterways shared by a large number of countries. The Bank has been more successful in helping to address transboundary disputes than in strengthening transboundary institutions. Work with its borrowers on transboundary aquifers is still in the early stages.

Strategic Issues

The Bank's complementary strategies for the water sector have been broadly appropriate. However, their application thus far has underemphasized some of the most difficult challenges set by the 2003 strategy, and this has left some needs unmet. The Bank's approach to water will face heightened challenges brought about by climate change, migration to coastal zones, and the declining quality of the water resources available to most major cities and industry in the coming decades. These will require some shifts in emphasis.

Water stress needs to be confronted systematically. At present there is no statistical relationship between the amount of Bank water lending to a country and that country's water stress. The issue for the Bank is finding an entry point and helping the most water-stressed countries put the pieces together so that water needs can become more cen-

tral to their development strategy. This is not to say that the Bank should stop providing support to water-rich countries, or that increasing lending to water-stressed countries is the only or even necessarily the best solution. The failure to meet human needs for water and sanitation has its roots in political, economic, social, and environmental issues. These are becoming more entwined and cannot be solved unless a broader range of actors gets involved.

The most water-stressed group consists of 45 countries (35 of them in Africa) that are not only water poor but also economically poor. Country Water Resource Assistance Strategies have helped to place water resource discussions more firmly in the context of economic development in the countries where they have been done. Including ministries of planning and finance in the dialogue is another critical step, as is expanding the calculation of economic benefits to increase countries' understanding of the economic importance of water.

Collaboration with other partners is particularly important, and it is likely to increase in importance as the Bank helps countries tackle water crises. This is true not only for water supply and sanitation, but also for water resources management in national and transboundary basins. Many of the problems described in this report are far too big for the Bank to tackle on its own.

Successful implementation of the Bank's Water Resources Sector Strategy will require a great deal of data on water resources, and implementation needs to give a higher priority to data gathering going forward. Data on all aspects of water and relevant socioeconomic conditions need to be more systematically collected and monitored. Data need to be used better within projects. For example, the collection and analysis of up-to-date groundwater data are more important now than ever and need to be taken on board more commonly than they have been.

Recommendations

Work with clients and partners to ensure that critical water issues are adequately addressed.

- Seek ways to support those countries that face the greatest water stress. The mid-term strategy implementation review should suggest a way to package tailored measures to help the Bank and other donors work with these clients to address the most urgent needs, which will become far more challenging as water supply becomes increasingly constrained in arid areas.

- Ensure that projects pay adequate attention to conserving groundwater and ensuring that the quantity extracted is sustainable.

- Find effective ways to help countries address coastal management issues.

- Help countries strengthen attention to sanitation.

Strengthen the supply and use of data on water to promote better understanding of the linkages among water, economic development, and project achievement.

- In project appraisal documents, routinely quantify the benefits of wastewater treatment, health improvements, and environmental restoration.

- Support more frequent and more thorough water monitoring of all sorts in client countries, particularly the most vulnerable ones, and help ensure that countries treat monitoring data as a public good and make those data broadly available.

- In the design of water resources management projects that support hydrological and meteorological monitoring systems, pay close attention to stakeholder participation, maintenance, and the appropriate choice of monitoring equipment and facilities.

- Systematically analyze whether environmental restoration will be essential for water-related objectives to be met in a particular setting.

Monitor demand management approaches to identify what is working and what is not working, and build on these lessons of experience.

- Clarify how to cover the cost of water service delivery in the absence of full cost recovery. To the extent that borrowers must cover the cost of water services out of general revenue, share the lessons of international experience with them so that they can allocate costs most effectively.

- Identify ways to more effectively use fees and tariffs to reduce water consumption.

- Carefully monitor and evaluate the experience with quotas as a means to modulate agricultural water use.

Endnotes

Management Response

1. Management notes that this evaluation covers the role of the private sector in the delivery of water supply and sanitation, in which the Bank has a role only to the extent that a public-private partnership is involved. Management notes that this evaluation does not address the specific role of IFC and MIGA, as the "Approach Paper: IEG Evaluation of Bank Group Support for Water" (IEG 2008b) indicated it would do.

Chapter 1

1. The state of world water has been characterized as a crisis by the United Nations and other international organizations as well as in the popular press. See, for example, DFID (2008), Lall and others (2008), and Rogers (2008).

2. According to EM-DAT: The International Disaster Database, Université Catholique de Louvain, Brussels, Belgium (www.em-dat.net).

3. The Dublin Principles are as follows: "1. Fresh water is a finite and vulnerable resource, essential to sustain life, development and the environment. 2. Water development and management should be based on a participatory approach, involving users, planners and policy-makers at all levels. 3. Women play a central part in the provision, management and safeguarding of water. 4. Water has an economic value in all its competing uses and should be recognized as an economic good" (Global Water Partnership, "Dublin Statements and Principles," www.gwpforum.org/servlet/PSP?iNodeID=1345).

4. The MDGs are not the first time that the international community has set ambitious targets for water. In the early 1980s the goal was "Water and Sanitation for All by 1990." That unmet goal was reset a decade later with the declaration of the Third Water Decade.

5. World Bank (2003a). The Sustainable Infrastructure Action Plan approved in 2007 further refined the relevant guidance.

Chapter 2

1. There were also 35 recipient-executed activities, 30 special financings, 7 guarantees, and 56 grants.

2. In addition, the International Finance Corporation (IFC) approved $752 million in financing for 45 projects in the water sector between 1997 and 2007 (not evaluated in this report; for additional information on the IFC portfolio, see appendix tables J.8–J.10). These projects were mainly for water utilities and large hydropower projects but also included a small number of water transportation and fisheries projects. Most of these projects (76 percent) were in the Asia and Latin America Regions.

3. The Bank-wide averages are larger because many non-water-related loans, such as Development Policy Loans, tend to be larger than those seen in projects with water activities.

4. The same type of overlap is possible between most categories; appendix J provides an overlap analysis.

5. The top 10 IDA borrowers for water, in order of amount committed, were India, China, Vietnam, Pakistan, Tanzania, Uganda, Ethiopia, Bangladesh, Nigeria, and Sri Lanka.

6. All projects prepare an Implementation Completion Report (ICR) when they close. Part of this process involves the assignation of performance ratings. These are reviewed and validated by IEG for quality control purposes. In addition, 25 percent of all projects are subject to field verification by IEG. If this leads to a ratings change, the revised rating is used in major IEG evaluations such as this one.

7. The overall water portfolio ratings lagged a percentage point behind the rest of the Bank portfolio in likely sustainability (67 versus 68 percent) and substantial institutional development impact (46 versus 47 percent) during the period 1997–2006, when IEG assigned such ratings. Borrower performance was slightly better than the Bank average, and Bank performance was about the same as for the Bank-wide portfolio. Annex J provides more detail on the ratings analysis.

8. IEG (2008b). The 2009 *Annual Review of Development Effectiveness* was not used in this analysis because it does not report five-year averages and does not focus on water. It does, however, show a small performance improvement from the 2003–05 period to the 2006–08 period.

9. Data for 2007 are not based on the evaluation portfolio but rather are from *Annual Review of Development Effectiveness 2008* (IEG 2008a).

10. Comparing open and closed projects provides an indication of whether the focus of Bank water projects has shifted, and comparing the percentage of ongoing projects in each subset with the percentage of ongoing projects for the water portfolio as a whole shows the overall relative extent of Bank attention.

11. The calculation of the index is described as follows: "WPI is calculated based on five components: Resources,

Access, Capacity, Use, and Environment. Each of these five indices is composed of sub-indices. Each sub-index is generated by a group of indicators. A country's sub-index value is evaluated by its position relative to the values in other countries, usually producing a value between 0 and 1 (values less than 0 are assigned 0 and values greater than 1 are assigned 1). The sub-indices are averaged and multiplied by 20 to produce one of the five, equally-weighted, component indices. Once all five component indices have been calculated, they are added together, producing a value between 0 and 100. This value is the water poverty index. The Resources component addresses the issue of water availability. It is comprised of internal water resources and external water flows indices. Less emphasis is put on external water flows than on internal water resources, as external resources are considered less reliable. Access is made up of indicators which address basic water and sanitation needs as well as the water requirements for agricultural food production. Capacity addresses several socio-economic factors that can impact water access. This index is composed of GDP [gross domestic product] per capita (adjusted for the currency's purchasing power), under-five mortality rate, the education index calculated by the United Nations Development Programme, and the Gini coefficient, which describes the distribution of income across a population. Use incorporates the water use, on a per capita basis, in the domestic, industrial, and agricultural sectors. Environment attempts to address not only water quality issues but also those issues of environmental governance. This index averages five component indices: water quality, water stress, regulation and management capacity, informational capacity, and biodiversity" Natural Environment Research Council, Centre for Ecology and Hydrology (2002). http://earthtrends.wri.org (accessed 10/19/09).

12 . As a group, water-poor countries borrow slightly more for water-related projects (38 percent of their total borrowing) than do non-water-poor countries (34 percent). When the analysis is repeated for dedicated projects only, the difference is even smaller: 8.7 percent of total lending in water-poor countries versus 8.9 percent in non-water-poor countries. The correlation coefficient between the share of borrowing for water and the WPI is 0.01, pointing to the virtual absence of a relationship.

Chapter 3

1. These problems include increasing prevalence of monocropping, erosion as a consequence of deforestation, and the myriad inappropriate agricultural practices that can cause a steep decline in the productivity of soils, not to mention the poor coverage and quality of agricultural support services.

2. Even in projects paying special attention to women and minorities, women's preferences were often not sought; see chapter 5, box 5.2.

3. *Water Encyclopedia,* "Artificial Recharge," www.water-encyclopedia.com/A-Bi/Artificial-Recharge.html.

4. The information available on the remaining project was not sufficient to allow a determination of whether the basin management organization had been established as planned.

5. Only early warning systems that used hydrological and meteorological monitoring systems were included.

6. This figure does not count hydropower projects that promote electricity conservation.

7. IEG is preparing an evaluation of economic analysis in projects for expected delivery in 2010.

8. Some projects in this category almost achieved the economic returns estimated at appraisal. The ERRs for these projects ranged between -0.1 and -44 percent, with a median of -7 percent.

9. The Bank's 2003 Water Resources Sector Strategy notes that "For decades there has been a yawning gap between simple economic principles (farmers should pay the full financial costs—operation and maintenance, rehabilitation, debt servicing on existing infrastructure—and the opportunity costs of water) and on-the-ground reality" (World Bank 2003b).

10. "Pricing and Subsidies," go.worldbank.org/SJBI3DFZW0, retrieved on October 7, 2009.

Chapter 4

1. Some regions have an annual cyclone season and are therefore counted more than once in this figure.

2. IEG (2007) found that 45 percent of loans responding to an emergency built roads. Of those projects that built roads in a postdisaster context, 98 percent financed the rehabilitation of disaster-damaged roads, while only 2 percent financed new construction of roads.

3. The impact of water scarcity and drought on energy systems was dramatically demonstrated in Brazil in 2001, when severe drought struck São Paulo state, which is heavily dependent upon hydropower. Unable to meet demand for electricity, state officials imposed electricity rationing selectively, according to the economic importance of particular industries and the jobs affected. The crisis cut the nation's gross domestic product by 2 percent, a loss of $20 billion.

4. "Drought Management," go.worldbank.org/YBDD5POI20, retrieved on October 7, 2009.

5. This definition is from the 10th International Rivers Symposium and International Environmental Flows Conference, held in Brisbane, Australia, on 3–6 September 2007. The conference was attended by more than 750 scientists, economists, engineers, resource managers, and policy makers from more than 50 countries (see www.riverfoundation.org.au/images/stories/pdfs/bnedeclaration.pdf).

6. Some projects called for putting in place laws to mandate EFAs (4 projects) or institutions to support EFAs (7 projects), or both.

7. According to the *Global Monitoring Report 2008* (IMF and World Bank 2008), China still tops the global list of industrial water polluters, emitting over 6 million kilograms of organic water pollutants a day; India is third, with emissions of over 1.5 million kilograms per day; in Brazil, regional and seasonal water scarcity in the northeast, in addition to water pollution, has created serious water quality problems, and the problem is growing. Pakistan faces water-related infections, and pollution-related problems, contributing to child mortality and contaminated water, remain a major risk factor for maternal health. Indonesia has one of the lowest rates of sewerage and sanitation coverage in Asia, which continues to cause widespread contamination of surface water and groundwater (Locussol 1997).

8. The Bank has also focused on hardware, building water and sanitation infrastructure, and drainage and irrigation systems. Yet little has been done to protect watersheds and wetlands, which are vital to the recharge of underground sources of water.

9. A 2007 evaluation from the Ministry of Foreign Affairs of Denmark found (DANIDA 2007) that "none of the countries [reviewed] have monitoring systems in place to measure WSS cost-effectiveness, by benchmarking unit costs for specific water and sanitation facilities or value for money studies. Therefore it is very difficult to measure sector efficiency. Improvements in sector monitoring fields are thus urgently needed." In addition, a 2004 evaluation undertaken by the Asian Development Bank's Operations Evaluation Department on the impact of water on the poor (ADB 2004, p. 21) states that, "Insufficient attention by ADB to project management and monitoring causes slow loan disbursements, adversely affecting project implementation and performance." The National University of Singapore's Institute for Water Policy undertook a review of 33 ADB water resources, water, and wastewater projects completed between 1995 and 2008. A total of 58 percent of the completion reports for these 33 projects mentioned the importance of monitoring. The need for monitoring systems ranked highest on the list of the 10 most important lessons to be learned (Aral and others 2009).

10. A World Bank Policy Research Working Paper (Kilgour and Dinar 1995) noted that of the 200 extant transboundary river basins, 145 are shared by 2 countries, 30 by 3, 9 by 4, and 13 by 5 or more—up to as many as 10.

11. A textual search of the Bank's entire water database for the 1997–2007 period identified 217 projects that included at least some discussion of rivers or lakes and related issues. However, a more in-depth review of the project documents revealed that only 174 projects made an identifiable attempt to undertake an activity relating to those river and lake issues; this is the group of projects analyzed as the "river and lake portfolio."

12. This figure also includes projects with coastal wetlands registered as Ramsar sites (wetlands protected under the 1971 Ramsar Convention).

13. Four hundred and ten project documents discussed wetlands, and 115 discussed the preservation or restoration of coastal mangroves (replanting). Some projects did both.

Chapter 5

1. The commitment amount was compiled based on the percentage within projects devoted to the WSS subsectors according the internal Bank database's coding.

2. The percentages for the five software activities sum to more than 57 percent of projects because some projects included more than one of the activities.

3. Basic sanitation, a subject of the MDG target, consists of human waste management at the household level. As defined by the World Health Organization (WHO), it is "the lowest-cost technology ensuring hygienic excreta and sullage disposal and a clean and healthful living environment both at home and in the neighborhood of users. Access to basic sanitation includes safety and privacy in the use of these services. Coverage is the proportion of people using improved sanitation facilities: public sewer connection; septic system connection; pour-flush latrine; simple pit latrine; ventilated improved pit latrine." WHO, "Water Sanitation and Health (WSH)," www.who.int/water_sanitation_health/mdg1/en/index.html, accessed October 22, 2009.

4. One latrine may serve many people; how many depends on the country. In countries such as Bolivia, Panama, and Paraguay, one latrine may be used by a small family of four; in rural Ghana, as many as 60 people tend to share one latrine.

5. These included, among others, the Guinea Third Water Supply and Sanitation Project (P001075), the Turkey Erzincan Earthquake Rehabilitation and Reconstruction Project (L3511), the Brazil Municipal and Low-Income Areas Project (Loan 2983), and the Abidjan Environmental Protection Project in Côte d'Ivoire (Loan 3155).

6. The health benefits of the Bank's WSS investments are rather obscure. Although half of the 117 WSS projects evaluated for the 2008 IEG health evaluation cited potential health benefits, and 89 percent financed infrastructure that plausibly could have improved health, only 1 in 10 listed improved health as an objective. Projects approved more recently (fiscal 2002–06) are even less likely to have been justified by health benefits, to have explicit health objectives, or to plan to collect health indicators. Only 14 water projects included health benefits in their economic analysis.

7. World Commission on Dams (2000). According to the International Commission on Large Dams, a "large" dam is one that is at least 15 meters high (from the foundation), or between 5 and 15 meters high and with a reservoir volume of more than 3 million cubic meters.

8. More guidance on this topic is provided by the World Bank (2009).

Chapter 6

1. For rural water supply, community management is the most notable institutional approach of the last two decades. This is discussed in the section on WUAs in the previous chapter and in appendix G.

2. The Turkey contract was abbreviated because the operator was dissolved and terminated because of unpredicted cost overruns.

3. According to the UNDP's *Human Development Report 2006: Beyond Scarcity*, "Major international companies such as Suez, the world's biggest water company, Veolia Environment and Thames Water are pulling back from concessions in developing countries, sometimes in the face of pressure from government and regulators. For example, Thames Water withdrew from the operation of a plant in China in 2004, two years after the Chinese government ruled that the rate of return was too high" (UNDP 2006, p. 92).

4. IEG (2008c). In the mid- and late 1990s, the Bank's approach to decentralization shifted from neutrality to strong partisan support for the decentralization of responsibilities. The current approach, however, is somewhat more cautious. Although decentralization is still seen as a potential contributor to public sector reform, it is no longer the only favored approach to institutional strengthening, and lack of capacity at the local level is now recognized as a major constraint to its broader application. The 2003 Water Resources Sector Strategy (World Bank 2003b), for instance, points out the substantial risks associated with decentralized water service provision and argues that optimal water resources management may need to be at the national or even international level.

5. See Global International Waters Assessment, "The GIWA LME Matrix," www.unep.org/dewa/giwa/publications/finalreport/back_cover.pdf, retrieved November 12, 2009.

6. However, as the project research regarding cost recovery in WSS found, most projects that attempt cost recovery are actually unsuccessful.

7. The deepest form of decentralization is the *devolution* of responsibilities, authority, and accountability to lower governmental levels, at least to some degree. Less extensive forms of decentralization include *delegation*, with the central government transferring some authority but the local government remaining accountable to the central government, and *deconcentration*, where some functions are delegated to the local level but the central government remains in charge (von Braun and Grote 2000). The latter two forms of decentralization usually transfer some administrative and political authority, but the fiscal dimension often remains at the central level, because decentralization of fiscal resources involves a significant loss of power for the central government.

8. This finding is also supported by an ordinary least squares regression of decentralization success on decentralization type and various other covariates.

9. See Ahmad and others (2005) for further details on the inefficiencies that can arise out of an incomplete decentralization process.

10. The 1993 strategy elaborates on the IWRM approach, stating that specific options for investment and development must consider the interrelations among different sources of water. Surface and groundwater resources are physically linked, and therefore their management and development should also be linked. Land and water management activities as well as issues of quantity and quality need to be integrated within basins or watersheds, so that upstream and downstream linkages are recognized and activities in one part of the river basin take into account their impact on other parts. Investments in infrastructure may displace people and disturb ecosystems. Thus, water resource assessments need to consider these cross-sectoral implications.

11. The first World Bank policy on international waterways was issued in 1956. Its current version (OP/BP 7.50, issued in 2001, currently under revision) covers rivers, canals, lakes, bays, gulfs, straits, and tributaries to those bodies of water, and it commits the Bank to facilitate cooperation and notify riparian countries of Bank projects on international waterways. The 1993 strategy also states that the incremental cost of actions taken by riparian states to protect international water resources and river basins will be financed within the framework of the GEF.

Relevant World Bank Policies

All policies available at:
http://www.worldbank.org/

OP 4.07—Water Resources Management (February 2000)
OP/BP 4.37—Safety of Dams (October 2001)
OP/BP 7.50—Projects on International Waterways
OP/BP 4.01—Environmental Assessment (January 1999)
OP/BP 4.02—Environmental Action Plans (February 2000)

Relevant World Bank Strategies

1993   *Water Resources Management Policy Paper*. Washington, DC: World Bank.

2000   *Cities in Transition: World Bank Urban and Local Government Strategy.* Washington, DC: World Bank.

2001   "Commitment to the Millennium Development Goals." World Bank, Washington, DC.

2001   *Making Sustainable Commitments: An Environment Strategy for the World Bank* (being updated). Washington, DC: World Bank.

2002   *Reaching the Rural Poor: A Renewed Strategy for Rural Development* [Irrigation]. Washington, DC: World Bank.

2003   *Water Resources Sector Strategy: Strategic Directions for World Bank Engagement.* Washington, DC: World Bank.

2005   Water Supply and Sanitation Business Plan (Water Supply and Sanitation).

2006   "Clean Energy and Development: Towards an Investment Framework." DC2006-002. Washington, DC, World Bank.

2008   "Global Food Crisis Response Program." http://www.worldbank.org/foodcrisis

2008   "Towards a Strategic Framework on Climate Change and Development for the World Bank Group: Concept and Issues Paper" (Ongoing). http://beta.worldbank.org/overview/strategic-framework-development-and-climate-change

2009   Hydropower Business Plan (Multi-purpose Use of Water Infrastructure for Energy Purposes).

2009   "The World Bank Urban and Local Government Strategy: Concept and Issues Note" (Ongoing). http://www.wburbanstrategy.org/urbanstrategy/sites/wburbanstrategy.org/files/Urban%20Strategy%20Concept%20Note%20FINAL.pdf

Other Publications

ADB (Asian Development Bank). 2004. "The Impact of Water on the Poor: Summary of an Impact Evaluation Study of Selected ADB Water Supply and Sanitation Projects." ADB Operations Evaluation Department, Manila.

AECOM International Development, Inc. 2009. "Inputs into the Preparation of the IBNET Blue Book. Final Report." Prepared for the International Bank for Reconstruction and Development, AECOM International Development Inc., Los Angeles, CA.

Ahmad, Junaid, Shantayan Devarajan, Stuti Khemani, and Shekhar Shah. 2005. "Decentralization and Service Delivery." World Bank Policy Research Working Paper 3603, World Bank, Washington, DC.

Ait Kadi, Mohamed. 1999. "Irrigation and Water Pricing Policy in Morocco's Large Scale Irrigation Projects." Paper presented at the World Bank Seminar on the Political Economy of Water Pricing, Washington, DC.

Aral, E., and others. 2009. "Water Management: Good Practices and Lessons Learned." Institute of Water Policy Research Report, Lee Kuan Yew School of Public Policy, Singapore.

Bardhan, Pranab. 2002. "Decentralization of Governance and Development." *Journal of Economic Perspectives* 16 (4): 185-205.

Behr, Peter. 2008. "Looming Water Crisis: Is the World Running out of Water?" *Congressional Quarterly Global Researcher* 2 (2): 27-56.

Biswas, A. K. 1999. "Management of International Waters: Opportunities and Constraints." *International Journal for Water Resources Development* 15 (4): 429–41.

Borger, Julian. 2007. "Darfur Conflict Heralds Era of Wars Triggered by Climate Change, UN Report Warns." *The Guardian*, June 23.

Briscoe, John. 1984. "Intervention Studies and the Definition of Dominant Transmission Routes." *American Journal of Epidemiology* 120 (3): 449–55.

Camdessus Panel. 2003. *Financing Water for All: Report of the World Panel on Financing Water Infra-structure.* World Water Council and the Global Water Partnership. Washington, DC: World Bank.

Clarke, Robin, and Jannet King. 2004. *The Water Atlas.* New York, NY: The New Press.

Comprehensive Assessment of Water Management in Agriculture. 2007. *Water for Food, Water for Life: A Comprehensive Assessment of Water Management in Agriculture.* London: Earthscan, and Colombo, Sri Lanka: International Water Management Institute.

Curtis, Val, and Sandy Cairncross. 2003. "Effect of Washing Hands with Soap on Diarrhea Risk in the Community: A Systematic Review." *Lancet Infectious Diseases* 3: 275–81.

DANIDA (Ministry of Foreign Affairs of Denmark). 2007. "Evaluation of Danish Support to Water Supply and Sanitation (1999–2005)." DANIDA, Copenhagen.

DFID (U.K. Department for International Development). 2008. *Water: An Increasingly Precious Resource. Sanitation: A Matter of Dignity.* London: DFID.

Dyson, Megan, Ger Bergkamp, John Scanlon, and IUCN Water and Nature Initiative. 2003. *Flow: The Essentials of Environmental Flows.* Gland, Switzerland: IUCN (International Union for the Conservation of Nature).

Faguet, Jean-Paul. 2004. "Does Decentralization Increase Government Responsiveness to Local Needs? Evidence from Bolivia." *Journal of Public Economics* 88: 867–93.

Falkenmark, M., A. Berntell, A. Jägerskog, J. Lundqvist, M. Matz, and H. Tropp. 2007. "On the Verge of a New Water Scarcity: A Call for Good Governance and Human Ingenuity." SIWI Policy Brief, Stockholm International Water Institute, Stockholm, Sweden.

Foster, Vivien. 2005. "Ten Years of Water Service Reform in Latin America: Toward an Anglo-French Model." Water Supply and Sanitation Sector Board Discussion Paper Series 3, World Bank, Washington, DC.

Galiani, Sebastian, Paul Gertler, and Ernesto Schargrodsky. 2003. "Water for Life: The Impact of the Privatization of Water Services on Child Mortality." Universidad de San Andres, University of California, Berkeley, and Universidad Torcuato Di Tella.

Gleditsch, Nils P., Ragnhild Nordås, and Idean Salehyan. 2007. "Climate Change, Migration, and Conflict." Coping with Crisis Working Paper Series, International Peace Academy, New York, New York.

Gleick, Peter, Heather Cooley, David Katz, and Emily Lee. 2006. *The World's Water 2006-2007: The Biennial Report on Freshwater Resources.* Washington, DC: Island Press.

Gleick, Peter, Meena Palaniappan, Mari Morikawa, Jason Morrison, and Heather Cooley. 2009. *The World's Water 2008–2009: The Biennial Report on Freshwater Resources.* Washington, DC: Island Press.

Global Water Partnership and Technical Advisory Committee. 2000. "Integrated Water Resources Management." TAC Background Papers 4, Stockholm, Sweden.

Groom, Eric, Jonathan Halpern, and David Ehrhardt. 2006. "Explanatory Notes on Key Topics in the Regulation of Water and Sanitation Services." Water Supply and Sani-

tation Sector Board Discussion Paper Series 6, World Bank, Washington, DC.

Hirji, Rafik, and Richard Davis. 2009. "Strategic Environmental Assessment: Improving Water Resources Governance and Decision Making." Water Sector Board Discussion Paper Series 12, World Bank, Washington, DC.

Hirji, Rafik, and Thomas Panella. 2003. "Evolving Policy Reforms and Experiences for Addressing Downstream Impacts in World Bank Water Resources Projects." *River Research and Applications* 19 (5): 667–81.

Hopkins, R., and David Satterthwaite. 2003. "An Alternative Perspective on WSS Services: Towns and the Urban/Rural Divide." In B. Appleton, ed., *Town Water Supply and Sanitation Companion Papers,* Vol. 3. Bank-Netherlands Water Partnership Project #043. Washington, DC: World Bank.

IEG (Independent Evaluation Group, World Bank Group). 2010. *Gender and Development: An Evaluation of Evaluation of World Bank Support, 2002–2008.* IEG Study Series. Washington, DC: World Bank.

———. 2008a. *Annual Review of Development Effectiveness: Shared Global Challenges.* IEG Study Series. Washington, DC: World Bank.

———. 2008b. "Approach Paper: IEG Evaluation of Bank Group Support for Water." http://www.worldbank.org/ieg

———. 2008c. *Decentralization in Client Countries: An Evaluation of World Bank Support, 1990–2007.* IEG Study Series. Washington, DC: World Bank.

———. 2007. "Development Actions and the Rising Incidence of Disasters." IEG Evaluation Brief Series, No. 4. IEG, Washington, DC.

———. 2006a. *Hazards of Nature, Risks to Development: An IEG Evaluation of World Bank Assistance for Natural Disasters.* IEG Study Series. Washington, DC: World Bank.

———. 2006b. *Water Management in Agriculture: Ten Years of World Bank Assistance, 1994–2004.* IEG Study Series. Washington, DC: World Bank.

———. 2002. *Bridging Troubled Waters: Assessing the World Bank Water Resources Strategy.* IEG Study Series. Washington, DC: World Bank.

———. 1996. "World Bank Lending for Large Dams: A Preliminary Review of Impacts." Report 15815, World Bank, Washington, DC.

IMF (International Monetary Fund) and World Bank. 2008. *Global Monitoring Report 2008: MDGs and the Environment: Agenda for Inclusive and Sustainable Development.* Washington, DC: World Bank.

———. 2007. "Clean Energy for Development Investment Framework: Progress Report of the World Bank Group Action Plan." Washington, DC, World Bank.

Intergovernmental Panel on Climate Change. 2007. "Summary for Policymakers." In *Climate Change 2007: The Physical Science Basis. Contribution of Working Group I to the Fourth Assessment Report of the Intergovernmental Panel on Climate Change,* ed. S. Solomon, D. Qin, M. Manning, Z. Chen, M. Marquis, K. B. Averyt, M. Tignor, and H. L. Miller. Cambridge, United Kingdom and New York: Cambridge University Press.

IUCN (International Union for Conservation of Nature). 2009. *An Integrated Wetland Assessment Toolkit. A Guide to Good Practice,* ed. Oliver Springate-Baginski, David Allen, and William Darwall. Gland, Switzerland: IUCN.

Jalan, Jyotsna, and Martin Ravallion. 2003. "Does Piped Water Reduce Diarrhea for Children in Rural India?" *Journal of Econometrics* 112 (January): 153–73.

Jette, Christian. 2005. "Democratic Decentralization and Poverty Reduction: The Bolivian Case." United Nations Development Programme, New York.

Kauffmann, C., and E. Perard. 2007. "Stocktaking of the Water and Sanitation Sector and Private Sector Involvement in Selected African Countries." Background Note for the Regional Roundtable on Strengthening Investment Climate Assessment and Reform in NEPAD Countries, Lusake, Zambia, November 27–28.

Kilgour, D. Marc, and Ariel Dinar. 1995. "Are Stable Agreements for Sharing International River Waters Now Possible?" Policy Research Working Paper WPS 1474, World Bank, Washington, DC.

Kokko, Hannu, and Oyj Vaisala. 2005. "Integrated Hydrometeorological Monitoring Solutions and Network Management." Paper presented at the 21st International Conference on Interactive Information Processing Systems (IIPS) for Meteorology, Oceanography, and Hydrology. Helsinki, Finland. Available at: http://ams.confex.com/ams/Annual2005/techprogram/paper_84648.htm.

Kolsky, Pete, Eddy Perez, Wouter Vandersypen, and Lene Odum Jensen. 2005. "Sanitation and Hygiene at the World Bank: An Analysis of Current Activities." *Water Supply and Sanitation Working Notes 6,* World Bank, Washington, DC.

Lall, Upmanu, Tanya Heikkila, Casey Brown, and Tobias Siegfried. 2008. "Water in the 21st Century: Defining the Elements of Global Crises and Potential Solutions." *Journal of International Affairs* 61 (2):1–22.

Litvack, Jennie, and Jessica Seddon. 1999. "Decentralization Briefing Notes." World Bank, Washington, DC.

Locussol, Alain. 1997. "Indonesia—Urban Water Supply Sector Policy Framework." Indonesia Discussion Paper Series 10, East Asia and Pacific Region Report 49944, World Bank, Washington, DC.

Marin, Philippe. 2009. *Public-Private Partnerships for Urban Water Utilities: A Review of Experiences in Developing Countries.* Washington, DC: World Bank.

Molden, David, ed. 2007. *Water for Food, Water for Life: Comprehensive Assessment of Water Management in Agriculture.* Colombo, Sri Lanka: International Water Management Institute.

Molle, François, and Jeremy Berkoff, eds. 2007. *Irrigation Water Pricing: The Gap between Theory and Practice.* Comprehensive Assessment of Water Management in Agriculture Series 4. Wallingford, United Kingdom: CABI.

Mwanza, Dennis. 2005. "Promoting Good Governance through Regulatory Frameworks in African Water Utilities." *Water Science & Technology* 51 (8): 71–79.

Natural Environment Research Council, Centre for Ecology and Hydrology. 2002. *The Water Poverty Index: International Comparisons.* Wallingford, United Kingdom: Centre for Ecology and Hydrology.

OECD-DAC (Organisation for Economic Co-operation and Development, Development Assistance Committee). 2004. *Lessons Learned on Donor Support to Decentralization and Local Governance.* DAC Evaluation Series. OECD, Paris.

Omar, Azfar, Satu Kähkönen, and Patrick Meagher. 2001. "Conditions for Effective Decentralized Governance: A Synthesis of Research Findings." IRIS Center, University of Maryland, College Park, MD.

Overbey, Lisa. 2008. "The Health Benefits of Water Supply and Sanitation Projects: A Review of the World Bank Lending Portfolio." IEG Working Paper 2008/1, Report No. 43207, Washington, DC.

Parker, Ronald, and Tauno Skytta. 2000. "Rural Water Projects: Lessons from OED Evaluations." IEG Working Paper Series, World Bank, Washington, DC.

Perard, Edouard. 2008. "Private Sector Participation and Regulatory Reform in Water Supply: The Southern Mediterranean Experience." Working Paper 265, OECD Development Center, Paris, France.

Pilgrim, Nick, Kevin Taylor, Sophie Tremolet, Ross Tyler, and Time Yates. 2003. "Business Planning for Town Water Services: Guidance Manual," Vol. 2. World Bank Working Paper 44725. Bank-Netherlands Water Partnership Project #043. Washington, DC.

Prud'homme, Rémy. 1995. "The Dangers of Decentralization." *World Bank Research Observer* 10 (2): 201–20.

Public-Private Infrastructure Advisory Facility, and World Bank. 2006. *Approaches to Private Participation in Water Services: A Toolkit.* Washington, DC.

Rees, Judith A. 2006. "Urban Water and Sanitation Services: An IWRM Approach." TEC Background Papers 11, Global Water Partnership, Stockholm, Sweden.

Revenga, Carmen, Jake Brunner, Norbert Henninger, Ken Kassem, and Richard Payne. 2000. *Pilot Analysis* of Global Ecosystems: Freshwater Systems. Washington, DC: World Resources Institute.

Rijsberman, Frank, interviewed by Kerry O'Brien. 2006. "Water Scarcity 'Due to Agriculture." Transcript. Australian Broadcasting Corp. (August 16). www.abc.net .au/7.30/content/2006/s1716766.htm.

Rogers, Peter. 2008. "Facing the Freshwater Crisis." Scientific American, July 23.

Sachs, Jeffrey, John McArthur, Guido Schmidt-Traub, Chandrika Bahadur, Michael Faye, and Margaret Kruk. 2004. "Millennium Development Goals Needs Assessments. Country Case Studies of Bangladesh, Cambodia, Ghana, Tanzania and Uganda." Working Paper, Millennium Project, New York.

Saghir, Jamal. 2004. "Hydropower: Beyond the Crossroads. Hydro 2004." PowerPoint presentation, World Bank Energy and Water Department, Porto, Portugal.

Salman, M. A. 2009. "The World Bank Policy for Projects on International Waterways. An Historical and Legal Analysis." Report 48741, World Bank, Washington, DC.

Scheierling, S. M., R. A. Young, and G. E. Cardon. 2006. "Public Subsidies for Water-Conserving Irrigation Investments: Hydrologic, Agronomic, and Economic Assessment." Water Resources Research 42 (3).

Schwartz, Klaas. 2008. "The New Public Management: The Future for Reforms in the African Water Supply and Sanitation Sector?" Utilities Policy 16: 49–58.

Shah, Anwar. 1998. "Balance, Accountability, and Responsiveness: Lessons about Decentralization." Policy Research Working Paper, World Bank, Washington, DC.

Sullivan, Caroline. 2003. "The Water Poverty Index: A New Tool for Prioritisation in Water Management." World Finance: 32–34.

———. 2002. "Calculating a Water Poverty Index." World Development 30: 1195–1210.

Sullivan, Caroline, and Jeremy Meigh. 2003. "The Water Poverty Index: Its Role in the Context of Poverty Alleviation." Water Policy 5:5.

Tanzi, Vito. 1996. "Fiscal Federalism and Decentralization: A Review of Some Efficiency and Macroeconomic Aspects." In Annual World Bank Conference on Development Economics, edited by Michael Bruno and Boris Pleskovic, pp. 295–316. Washington, DC: World Bank.

UNDP (United Nations Development Programme). 2006. Human Development Report 2006: Beyond Scarcity: Power, Poverty and the Global Water Crisis. New York, NY: UNDP.

UNEP (United Nations Environment Programme). 1999. "Conceptual Framework and Planning for Integrated Coastal Area and River Basin Management: Priority Actions Programme." UNEP, Nairobi.

UNICEF (United Nations Children's Fund) and WHO (World Health Organization). 2006. Meeting the MDG Drinking Water and Sanitation Target: The Urban and Rural Challenge of the Decade. New York and Geneva.

UN Millennium Project Task Force on Water and Sanitation. 2008. Health, Dignity, and Development: What Will It Take? London and Sterling, VA: Earthscan Publications.

United Nations. 2008a. Trends in Sustainable Development: Agriculture, Rural Development, and, Desertification and Drought. New York, NY: United Nations.

———. 2008b. "UN Millennium Project." http://www. unmillenniumproject.org/reports/tf_watersanita tion.htm. Retrieved on November 14, 2008.

Von Braun, Joachim, and Ulrich Grote. 2000. "Does Decentralization Serve the Poor?" Paper presented at the International Monetary Fund Conference on Fiscal Decentralization, Washington, DC, November 20–21.

World Bank. 2009. "Directions in Hydropower: Scaling Up for Development." Water Working Notes 21, World Bank, Washington, DC.

———. 2008a. "The Economic Impacts of Poor Sanitation." World Bank, go.worldbank.org/TF5BZG8GL0.

———. 2008b. "General Principles for Subsidies." World Bank, go.worldbank.org/3LLTDB4H80 (accessed October 15, 2008).

———. 2008c. "General Principles for Subsidy Design." World Bank, go.worldbank.org/3LLTDB4H80 (accessed November 23, 2008).

———. 2008d. "Subsidy Mechanisms." World Bank, go .worldbank.org/1A40DK4KO0 (accessed November 19, 2008).

———. 2008e. "Watershed Management Approaches, Policies, and Operations: Lessons for Scaling Up." Water Sector Board Discussion Paper Series 11, World Bank, Washington, DC.

———. 2007a. "Emerging Public-Private Partnerships in Irrigation Development and Management." Water Sector Board Discussion Paper Series 10, World Bank, Washington, DC.

———. 2007b. World Development Report 2007: Development and the Next Generation. Washington, DC: World Bank.

———. 2006a. Reengaging in Agricultural Water Management: Challenges and Options. Directions in Development Series 35520. Washington, DC: World Bank.

———. 2006b. Scaling Up Marine Management: The Role of Marine Protected Areas. Report 36635–GLB. Washington, DC: World Bank.

———. 2006c. "Tanzanian Water Resources Assistance Strategy: Improving Water Security for Sustaining Livelihoods and Growth." World Bank, Washington, DC.

———. 2005a. "The Principle of Managing Water Resources at the Lowest Appropriate Level—When and Why Does It (Not) Work in Practice." World Bank, Washington, DC.

———. 2005b. Water, Electricity, and the Poor. Washington, DC: World Bank.

———. 2003a. "Implementing the World Bank Group Infrastructure Action Plan (with Special Emphasis on Follow-up to the Recommendations of the World Panel on Financing Water Infrastructure)." Report 30618, World Bank, Washington, DC.

———. 2003b. "Water Resources Sector Strategy: Strategic Directions for World Bank Engagement." World Bank, Washington, DC.

———. 1998. "International Watercourses: Enhancing Cooperation and Managing Conflict." World Bank Technical Paper 414, World Bank, Washington, DC.

———. 1997. "Facilitating Private Involvement in Infrastructure." An action program, paper presented at the 56th meeting of the Development Committee, Hong Kong, 22 September.

———. 1993. *Water Resources Mangement Policy Paper.* Washington, DC: World Bank.

World Bank and PPIAF (Public-Private Infrastructure Advisory Facility). 2009. "Building Bridges: China's Growing Role as Infrastructure Financier for Sub-Saharan Africa." Trends and Policy Options 5, World Bank, Washington, DC.

World Bank, IFC (International Finance Corporation) and MIGA (Multilateral Investment Guarantee Agency. 2009. *Directions in Hydropower.* Washington, DC: World Bank.

World Commission on Dams. 2000. *Dams and Development: A New Framework for Decision-Making: The Report of the World Commission on Dams.* London and Sterling, VA: Earthscan Publications.

World Panel on Financing Water Infrastructure. 2003. *Financing Water for All: Report of the World Panel on Financing Water Infrastructure.* Washington, DC: World Bank.

Zandvliet, Luc, and Mary B. Anderson. 2009. *Getting it Right: Making Corporate-Community Relations Work.* Sheffield, United Kingdom: Greenleaf Publishing.

Photographs

www.ingramcontent.com/pod-product-compliance
Lightning Source LLC
Chambersburg PA
CBHW080243270326
41926CB00020B/4353